Effects of Welding on Health

A literature survey and evaluation to establish the state-of-the-art and to point the directions for future research to understand and improve the occupational health of welding personnel

Research performed at the Franklin Research Center under contract with the American Welding Society and supported by industry contributions

Prepared for:

SAFETY AND HEALTH COMMITTEE
AMERICAN WELDING SOCIETY
550 N.W. LeJeune Road, Miami, Florida 33126

Library of Congress Number: 79-52672
International Standard Book Number: 0-87171-180-X

American Welding Society, 550 N.W. LeJeune Road, Miami, FL 33126

Printed in the United States of America

Contents

Personnel

Authors of the report by Franklin Research Center were:

Jon E. Villaume
Khizar Wasti
Deborah Liss-Suter
Shu-hsi Hsiao

AWS Research Committee

J.F. Hinrichs, Chairman	A.O. Smith Corporation
F.Y. Speight, Secretary	American Welding Society
R.E. Kinser	Caterpillar Tractor Co.
P. Ramsey	A.O. Smith Corporation
P.C. Krueger	Airco Welding Products
A. Lesnewich	Airco Welding Products
N. Ward	Caterpillar Tractor Co.

AWS Research Finance Committee

E.B. Scripture, Chairman	Teledyne McKay
F.Y. Speight, Secretary	American Welding Society
W.B. Howell	Consultant
R.E. Bruggeman	Caterpillar Tractor Co.
R.H. Schwegman	General Electric Co.
E.H. Hunt	Allegheny Ludlum Industries

Acknowledgments

The American Welding Society gratefully acknowledges the time and effort expended by the members of the Research Committee and the Research Finance Committee and the financial support of the program by industry contributions.

Supporting Organizations

Alloy Rods
Division of Allegheny Ludlum Steel Company
Westinghouse Electric Company
Airco, Inc.
Miller Electric Company
The Heil Company
United States Steel Corporation
Union Carbide Corporation
Babcock and Wilcox Company
General Electric Company
E.I. DuPont de Nemours & Company
Teledyne McKay, Inc.
J.I. Case Company
Lincoln Electric Company
Huntington Alloys, Inc.
Stoody Company
AWS Detroit Section
Duke Power Company
Sandvik, Inc.

In addition, 21 other organizations made contributions to help support the program during the period January 1978 – April 1, 1979.

Preface

This literature review has been prepared for the Safety and Health Committee of the American Welding Society to provide an assessment of current knowledge of the effects of welding on health, as well as to aid in the formulation of a research program in this area, as part of an ongoing program sponsored by the Committee. Previous work has included studies of fumes and gases, radiation, and noise generated during various forms of arc welding. Conclusions based on this review and recommendations for further research are presented in the introductory portions of the report. Appendix A of this report reviews AWS sponsored studies and others in order to clarify the occupational exposures that are summarized in Chapter 1. Chapters 2 and 3 contain information related to the effects of exposure to byproducts of welding operations on humans and on laboratory animals. Chapter 4 covers studies of the mutagenicity of welding fume condensate.

Introduction

According to census figures (Ref. 1), there were an estimated 572,000 welders and cutters in 1970. Sosnin (Ref. 2) has estimated that today there are close to one million individuals whose occupation requires a substantial amount of welding; this latter estimate includes pipefitters and mechanics who view welding only as one of the tools of their trade. Therefore, the health and safety of this large occupational group is an important concern.

Welders are not a homogeneous group. They work under a variety of conditions: outdoors, indoors in open as well as confined spaces, underwater, and above ground on construction sites. They also utilize a large number of welding and cutting processes. However, many of these have in common the production of fumes, gases, radiation, and other potentially harmful agents. The review that follows attempts to assess the effects of these agents on the health of welders.

The discussion concentrates on health effects due directly to the welding process. It does not attempt to treat safety problems, nor does it cover in any detail the accompanying hazards that, although often found in the welding environment, are not created directly by the welding process.

Occupational health depends upon the conditions and exposures received in the workers' environment. In Chapter 1 of this report, the physical and chemical agents to which welders may be exposed and the conditions of their generation are summarized; details are provided in Appendix A. In Chapter 2, the effects of this exposure on welders and others in the environment are reviewed. Since welders are exposed to a number of factors simultaneously, the chapter is subdivided by organ system, not by agent. Effects on the respiratory system, skin, eyes, and internal organs are considered. Chapter 3 presents the results of studies in experimental animals; the review concludes with a discussion of special studies of the mutagenicity of welding fumes.

Executive Summary

Most commonly used welding processes emit fumes, gases, electromagnetic radiation, and noise as byproducts of their operation. During welding, workers are potentially exposed to all of these agents. The fumes are chemically very complex, arising primarily from the filler metals and any electrode coatings or cores. The potential exposure varies with the process and welding conditions employed.

Numerous reports of the effects of welding on health have been published. Many cases of acute poisoning, due to excess exposure to one or more gases and fumes, have been documented. However, other than lung involvement, few chronic effects have been attributed to welding, and, in almost all the studies of chronic effects, welding of ferrous metals was involved. This does not necessarily mean that welding ferrous or any other metals presents no routine health risks. Frequently, published studies have not been appropriately designed to detect the effects of chronic exposure and have not adequately examined special groups (for example, welders of aluminum or stainless steel) with unique exposure factors. Conclusions resulting from this review are presented briefly in the following sections.

Acute Poisoning

Exposure to ozone generated during gas shielded arc welding, especially of aluminum, may produce irritation and inflammation of the respiratory tract, excess fluid and hemorrhage in the lungs, and sometimes headache, lethargy, and eye irritation. Severe cases could be fatal. Ozone is generated at a distance from the arc by the action of the emitted ultraviolet radiation ($<$240 nm) on atmospheric oxygen. Nitrogen oxides, which are produced in welding arcs and flames by thermal oxidation of atmospheric nitrogen, produce similar respiratory tract effects.

Metal fume fever occurs in welders who inhale zinc-containing fumes, although other fume components may also produce the symptoms of this condition; for example, a metallic taste, chills, thirst, fever, muscle aches, chest soreness, fatigue, gastrointestinal pain, headache, nausea, and vomiting. The symptoms usually subside within 1 to 3 days of exposure without residual effect.

Chronic Lung Conditions

Protracted inhalation of welding fume particles leads to their accumulation in the lungs, a condition seen as dense areas on chest x-ray films. The severity of changes in lung x-rays is proportional to the length of welding experience; however, the changes seen in such x-rays are not necessarily associated with reduced lung function or disease.

In general, welders of ferrous materials have been found to have an increased frequency of respiratory symptoms such as chronic bronchitis. No conclusive studies have been made to determine whether reduced lung function, emphysema, or other chronic respiratory conditions generally occur more frequently in these groups than in nonwelders.

Significant levels of ozone are produced in certain gas shielded welding operations. Lengthy exposure of experimental animals to this gas has induced the formation of fibrous tissue and caused emphysema-like and other effects in the lungs. No studies of the long-term effects of this gas on welding populations exposed to it have been undertaken.

Studies of the effects of welding fumes on experimental animals have indicated excess deposition of fibrous tissue in the lungs when a variety of metals were welded. However, in these experiments, fume concentrations were always very high, and experimental conditions were incompletely described; therefore, it is difficult to relate the importance of these findings to the welding experience.

Eye

Ultraviolet (>170 nm), visible, and infrared radiation are emitted by the welding arc. Welders not wearing eye protection and others in the vicinity of the arc are at risk to the effects of this radiation.

Ultraviolet radiation produces the condition known as "arc eye," an acute inflammation of the external structures of the eye; the symptoms disappear within 1 to 2 days. Infrared radiation penetrates the interior of the eye and can cause burns on the retina.

There is no evidence that cataracts occur more frequently in arc welders than in nonwelders; the results of studies designed to detect cataracts in welders have yet to be published.

Skin

Exposed skin is susceptible to the effects of ultraviolet radiation from the arc (for example, erythema) and of any fume components capable of skin sensitization or irritation. Chromium compounds, which occur in stainless steel welding fumes, are a frequent cause of dermatitis.

Cardiovascular Disease

Electrocardiograms and blood pressures have been measured in several studies. In most of these, electrocardiograms of welders did not differ significantly from control groups of nonwelders, and welders have been shown to have normal or slightly lower blood pressure than the control groups.

Carbon monoxide is generated from carbon dioxide in the gas shielded welding processes. It combines avidly with hemoglobin, reducing the oxygen-carrying capacity of the blood. Exposure to carbon monoxide may present an added health risk to those with heart disease. No studies of heart disease in welders performing carbon dioxide shielded welding have been documented.

Nervous System and Other Organs

Lead and manganese can occur in the fumes of certain welding operations; both are toxic to the nervous system. Except for those attributable to these two substances, no reliable reports of effects of welding exposure to the nervous system have been published.

There is no significant evidence that welders have a higher incidence of other chronic diseases.

Carcinogenicity

Based upon available information, there does not appear to be any evidence that exposure to welding fumes and gases induces lung or other forms of cancer. Further study is required of groups exposed to specific kinds of fumes. The welding arc emits ultraviolet radiation of wavelengths that produce skin tumors in animals, but no study of this effect in welders has been reported.

Technical Summary

The welding exposure is unique. There is no material from any other source directly comparable to the composition and structure of welding fumes. Exposure to byproduct gases accompanies fume exposure; therefore, ability to extrapolate from exposures received by workers in other industries to those in the welding environment is limited.

Welders are potentially exposed to a large number of chemical and physical agents. Several questions of the effects of welding on health are apparent:

1. Do welders have a higher incidence of chronic respiratory tract disease than nonwelders?

2. Do they have a higher incidence of respiratory tract cancer or other malignancies?

3. Does the ingestion and inhalation of welding fumes and gases result in any other systemic diseases?

Since many different materials and processes are used, these questions should be asked for specific subgroups of welders. In available studies, adequate data to make these distinctions often have not been reported, and any conclusions made apply to welding in general. However, it should be noted that most welding involves ferrous-based metals.

Chronic Respiratory Disease

Protracted inhalation of welding fume particles leads to their accumulation in the lungs. Deposits of these fume particles in the lungs have been noted as dense areas on chest x-ray films. The severity of the changes in the x-rays has been shown to be proportional to the duration of welding exposure (Refs. 3, 4, and 5). However, this evidence of siderosis and pneumoconiosis is not necessarily associated with clinical evidence of lung disease.

Ten major epidemiological studies have contributed to the assessment of risk of chronic restrictive or obstructive lung disease or both in welders. All studies attempted to correct for the confounding effects of smoking. Most obtained previous occupational or medical histories or both (Refs. 3, and 6 through 15). However, in a number of studies (Refs. 3, 8, 9, 11 through 14, 16, and 17), the subjects were shipyard welders in whom concomitant exposure to asbestos is likely; in only one case (Ref. 8) was an attempt made to correct for this confounding effect. Although SMAW or FCAW or both were used by welders in all of the subject populations, descriptive details of the exposure varied between studies and were generally inadequate. All studies were published during 1964 or later except for one (Ref. 3) published in 1947; the latter may not be applicable to current conditions.

The most important criticism of all of these studies is that they were of the "point prevalence" or "cross-sectional" type. In this design, workers at a plant or plants at a particular point in time are studied. Serious bias may enter, because workers who are ill, have died, or have taken up other employment for health-related reasons are omitted. Point prevalence studies are likely to detect only subclinical disease.

Of nine studies that investigated respiratory symptoms, seven (Refs. 3, 7, 9, 11, and 15 through 17) reported an increased prevalence in welders compared to a reference group; in two (Refs. 6, 10, and 12 through 14), no significant difference was found. In only two (Refs. 9 and 16) could a pulmonary function deficit be reasonably attributed to the welding experience; this deficit was stated to be subclinical in one study (Ref. 9) and was probably subclinical in the second. Thus, from these studies, it appears that exposure to the welding fumes or gases or both caused irritation to the respiratory tract, but not of a sufficient degree to produce significant pulmonary dysfunction (Ref. 18) among current workers. However, the strength of these conclusions is seriously weakened by the bias inherent in point prevalence surveys.

There have been studies in which laboratory animals inhaled welding fumes and gases, or fume concentrate was administered to them intratracheally. Soviet and other Eastern European investigators have observed evidence of pulmonary fibrosis in rats (Refs. 19 through 26) and guinea pigs (Ref. 27) that usually involved experimentation with covered electrodes. It is not possible to implicate a particular agent or agents or to evaluate the importance of these findings because of deficiencies in these experiments, such as failure to report one or more of the following: electrode composition, fume level, fume composition, concentration of gases, and exposure schedule of the animal populations; in addition, dose-response relationships were not investigated.

The pulmonary hazard potential varies with the process and metals being welded. In most situations, fume production is the major source of atmospheric contamination; however, in certain gas shielded welding operations, ozone can be produced in significant amounts. The highest rate of ozone production occurs during GMAW (argon) of aluminum and its alloys (Ref. 28). Significant quantities are also produced in other GMAW and GTAW operations. Because the ozone is not produced in the arc, but is generated over a distance by the action of arc-emitted UV radiation on atmospheric oxygen, ozone removal is difficult. Exposure to welding radiation-produced ozone represents an acute respiratory hazard. The effects of chronic exposure to low levels of ozone in humans are unknown, but, in experimental animals, emphysematous and fibrotic and other pathologic pulmonary changes have been observed (Ref. 29).

Cancer

Epidemiological studies of cancer risk have only incidentally included welders (Refs. 30 through 35). Examination of a mixed group of welders was a specific goal in only one study (Refs. 34 and 35). This was a prospective study of lung cancer in 14 different occupational groups. Other studies involving welders include: a survey by NIOSH (Ref. 30) to examine causes of death in Washington state males; a cohort mortality study in Dow Chemical Company employees including maintenance welders (Ref. 31); a study of lung cancer deaths and cases in Los Angeles County (Ref. 32); and a case-control study of lung cancer patients in 11 California hospitals (Ref. 33). In three of these studies, no correction was made for the confounding effect of smoking (Refs. 30 through 32).

The results of three of the studies (Refs. 30, 32, and 33) indicated that further study of lung cancer in welders was warranted; but, in two of these, neither the effects of smoking nor occupational exposure was examined, and the population of welders was very small in the third. In the most reliable study (Refs. 34 and 35), an increased risk of lung cancer was not observed. Thus, there is no substantial evidence to indicate that exposure to welding byproducts causes lung cancer; reports of well designed epidemiological studies are not yet available.

Significant evidence is not available from epidemiological studies to indicate whether welders are exposed to an increased cancer risk for organs other than lungs. No reliable study of the carcinogenicity of welding fumes and gases in experimental animals has yet been reported.

Eye

Another major concern is effects to the eyes. Injury may be caused by ultraviolet radiation or flying particles, and irritation may be produced by certain fumes and gases (Ref. 36), particularly ozone (Ref. 37). In one Swedish shipyard in 1976, 4,000 welders had 11,000 eye accidents (Ref. 38). In 1977, 7,000 eye injuries in 3,000 welders were reported in another Swedish shipyard, 30 percent of which were due to UV radiation (Ref. 39). Ross (Ref. 40) reported 459 non-lost-time injuries among 400 heavy engineering welders in 1971; 43 percent involved the eye, one third of which were classified as arc eye.

A potential hazard is welding arc radiation. Ultraviolet, visible, and infrared radiation are produced by the welding arc. Welders wearing no eye protection are at risk when the arc is accidentally struck. Assistants and supervisors not adequately protected are also at risk.

Arc eye (or keratoconjunctivitis) is an acute, self-limited irritation and inflammation of the superficial structures of the eye (Refs. 41 and 42), resulting from cumulative unprotected exposure to the UV radiation from the arc (Ref. 43). Cascini (Ref. 44) noted that in 20 cases exposure to the arc lasted 0.33 to 2 hours. The symptoms disappear 22 to 52 hours after the exposure, and no permanent injury appears to remain (Refs. 42 and 45).

Although the eye is not penetrated by the ultraviolet portion of the spectrum, inner structures are affected by wavelengths of 400 to 1400 nm. Radiation in this region of the spectrum can cause lenticular cataracts and chorioretinal burns (Refs. 41 and 46).

Historically, a high incidence of cataracts has been observed in glass blowers and foundrymen after many years of unprotected exposure to radiation emitted by molten glass and metal, respectively (Ref. 41). Massive doses of IR radiation were required to produce lenticular damage in rabbits (Jacobsen et al., 1963, as cited in Ref. 46). This contrasts with the infrequent exposure in modern welders, which, except in cases of gross negligence, is of short duration.

In 1967, Edbrooke and Edwards (Ref. 46) reviewed available evidence and concluded that welders did not develop cataracts more frequently than the general population; unfortunately, no well documented epidemiological study in welders has been published. Although quite rare, cases of retinal damage after unprotected exposure to arc radiation have been reported (Ref. 47).

The distant and near vision of welders does not appear to be significantly different from that of nonwelders, based upon examinations of heavy engineering (Ref. 148) and shipbuilding welders (Refs. 3 and 48) and control groups. However, as in other studies of the eye, these are cross-sectional rather than cohort studies.

Other Organs and Systems

Cardiovascular System

No unusual EKG findings were observed among a mixed group of 402 welders who had at least five years of welding experience and had not been exposed to other pulmonary occupational hazards (Ref. 49). Likewise, EKG abnormalities in 35 shipbuilding welders did not occur more frequently than in a control group (Ref. 3). However, Italian investigators claim that the ratio of the P wave duration to the length of the PR segment was significantly higher in 58 steel arc welders than in controls (Refs. 50 and 51). The EKG should be monitored in future studies.

The blood pressure measurements of welders do not differ from those of nonwelders (Refs. 48 and 52) nor do they have lower blood pressure (Refs. 3, 53, and 54). The latter is possibly due to the fact that welders may be more physically fit than persons in sedentary jobs.

The production of carbon monoxide from carbon dioxide used in gas shielded welding operations and by the oxyacetylene flame is noteworthy. Carbon monoxide avidly combines with hemoglobin to reduce the oxygen-carrying capacity of the blood. Exposure to 100 ppm for 4 hours shortened the time of onset of exercise pain in patients with ischemic heart disease and altered their EKG (Ref. 55). It has been shown to aggravate peripheral arterial disease (Ref. 56) and lowered the threshold for ventricular fibrillation in dogs and monkeys with experimentally-induced heart disease (Ref. 57).

Blood

In 1976, Ross and Hewitt (Ref. 52) found no significant differences in hemoglobin levels between 350 heavy engineering welders and a group of controls. In 1947, Dreessen (Ref. 3) noted no significant differences in hemoglobin levels between 3,000 shipbuilding welders (including both sexes and blacks and whites) and nonwelders. Likewise, in 1964, Marchand et al. (Ref. 49) reported normal hemograms in 402 welders with varied welding experiences, and Schuler et al. (Ref. 58) found no anemia in 23 welders from Santiago, Chile. Thus, under the conditions of the above studies, welding does not appear to affect hemoglobin levels.

Although leukocytosis is reported to be a common finding in metal fume fever (Ref. 59), Chmielewski et al. (Ref. 60) reported normal white blood cell counts in 20 welders before and after they worked on galvanized metal in confined spaces. Also, the leukocyte counts were not different from those of controls among heavy engineering welders (Ross and Hewitt, Ref. 52). Data on other blood parameters are unavailable.

Nervous System

Potential exposure to manganese occurs whenever this metal is used in electrode coatings and cores or in electrode wire. Potential lead exposure occurs during welding and cutting of any metal coated with lead-based paint. Both are poisonous to the nervous system. Information on effects to the nervous system is unavailable, except for those effects attributable to manganese or lead.

Gastrointestinal Tract

Symptoms of metal fume fever include diarrhea, nausea, and abdominal pain (Ref. 60). Chromate, which may be generated in stainless steel welding fumes, is an irritant to mucosal tissue (Ref. 61). Although Dreessen et al. (Ref. 3) noted no increase in gastrointestinal complaints in shipyard welders compared to controls, Rozera (Ref. 62), in a study of 620 Italian metallurgical and metal machining welders, found increased morbidity from ulcerative and other digestive system diseases in them compared to other workers in the industry. Significant corroborating data are lacking.

Other Organs

Except for a few cases of cadmium poisoning, information on effects to the kidneys is unavailable. No reliable reports of adverse effects on liver function were located. One report on a small group of 25 foundry welders revealed no clinical evidence of endocrine dysfunction or significant differences in plasma cortisol, urinary corticosteroids, plasma and urine epinephrine or norephinephrine, compared to 10 controls (Ref. 63).

Welders may be exposed to fluoride dust and vapors from certain FCAW and GMAW operations and SAW fluxes. Chronic fluorosis is a syndrome characterized by increased density of bones and ligaments due to fluoride deposition. However, no data are available that identify a relationship between exposure to fluoride-containing fumes and disorders of bones or ligaments or both (Ref. 64).

Recommendations

Available data are not adequate to reach many conclusions regarding health effects, if any, of the welding experience. Recommendations to fill in the most important information gaps are listed as items 1 through 4 below. In addition, potential health considerations should become an integral part of the development of new welding products; items 5 and 6 are suggested to aid in studying health hazard potential during the development stage.

1. Prospective epidemiological study of chronic lung disease. All of the studies reported to date are of the "cross-sectional" type and are subject to bias from lost workers (Ref. 65). The study should start with a cohort of welders, all employed at the start of the study in the same or similar welding environments; a control group of the same size should be included. Previous occupational exposures, smoking habits, and medical histories should be obtained. The groups should then be followed for a defined period with periodic observation, using symptom questionnaires, chest x-rays, and simple pulmonary function tests. All illnesses and hospitalizations should be carefully documented. All medical examinations and analyses should be performed without knowledge of occupation. Careful environmental monitoring should be pursued so that specific cause-and-effect relationships, if any, can be elucidated (Ref. 18). Increased information can be gained by combining the subject group with an historical cohort, starting 3 to 5 years in the past.

This study would also provide an excellent opportunity to monitor EKG, clinical chemistry, and blood metal levels. The size of the welder and reference groups to be chosen, as well as the duration of the study, depend upon the particular welding environment being examined.

Two milieus should be examined. First, a study of SMA or FCA welding of ferrous alloys is warranted. Such welding is commercially important, and, therefore, large numbers of individuals are potentially exposed. Results of earlier cross-sectional studies indicated that a large population and a long study period (5 years or more) may be required to pick up any effects when good industrial hygiene practices are observed (Ref. 18).

The second group recommended for prospective monitoring are welders using gas metal arc processes where ozone generation is significant. If the atmosphere in this type of welding environment represents a significant health risk, it may be detectable with a smaller sample.

2. Epidemiologic studies of carcinogenic potential. No reliable studies are available that assess the carcinogenic potential of any particular welding situation. Investigation of carcinogenicity will require selection of a cohort of welders and nonwelding controls, in existence far enough in the past that any cancer would have had enough time to develop. A reconstruction of past industrial hygiene measures at the plant under investigation and retrospective collection of relevant personal and occupational information on the subjects (or next of kin) should be undertaken. The incidence and mortality patterns in the exposed and control groups must then be ascertained. Death and morbidity data on other chronic diseases will also be concomitantly obtained.

An increased risk of respiratory tract cancer has been associated with exposure to certain chromates (Ref. 66) and nickel compounds (Ref. 67). Therefore, welders of stainless steel and other chromium and nickel alloys should be studied. The National Cancer Institute is currently sponsoring a study of welders of stainless steel (Ref. 68).

Shielded metal arc or flux cored arc welding of ferrous materials or both account for a major portion of today's welding operations. Therefore, effects of exposure to this milieu in the absence of chromium and nickel (for example, in mild steel welding) should be considered if stainless steel welding is found to be associated with an increased carcinogenic risk.

Results from an epidemiologic study of this design can provide information only on the carcinogenic risk under industrial hygiene conditions and processes that existed 15 to 30 years ago. Therefore, any cancer epidemiology data should be supplemented with information on current exposure conditions (see item 6).

3. Prospective study of eye injuries. Although neither heat cataracts nor retinal injury have been reported to be a problem in welders, published supporting data are poor. Cohorts of current welders should be chosen, and they should undergo periodic eye examination for a length of time sufficient to detect cataract development. If possible, workers who seek other employment should also be followed, since attrition may be due to eye problems. This study might be undertaken independently or as part of item 1.

4. Design of employee records forms. Model forms should be developed to facilitate collecting and centralizing all job description, environmental monitoring, personnel, and other information necessary to monitor the health of welder populations. These should be made available to all firms employing large numbers of welders.

5. Screening of fume for fibrogenic potential. Fume from SMAW and FCAW is a complex mixture composed of unoxidized and oxidized metal from the electrode and material originating from the coating and core, respectively. Although significant evidence of restrictive lung disease was not uncovered in the limited cross-sectional studies discussed above, experiments in animals indicate possible fibrogenic potential.

Silicates constitute a significant portion of many coatings and cores, releasing silica or silicates or both into the fumes when these electrodes are burned. Another source of silica is in the oxidation of alloying elemental silicon. Crystalline forms of silica, which are known to produce pulmonary fibrosis (Ref. 69), have not been found in welding fumes (Refs. 70 and 71). However, even amorphous silica (Refs. 72 through 75, and 305), such as that formed by evaporation from liquid silica (Ref. 72), or amorphous hydrated silica fibrils obtained by acid treatment of chrysotile (Ref. 73) may have fibrogenic potential. Therefore, consideration should be given to fibrogenic potential when developing new electrodes.

Ideally, electrodes should be screened for the fibrogenic potential of their fumes, using a rapid and inexpensive assay. An *in vitro* assay meeting these requirements is the alveolar macrophage test (Refs. 73 and 74). This assay is in the developmental stage and should be first examined for its utility in screening fume condensate.

If the test is validated, electrodes with standard or "typical" coating and core compositions should be developed for such a program, as it is important to be able to correlate biological effects with coating or core compositions. Results on "typical" electrodes can then be made available to aid manufacturers in coating and core formulation.

Based on results of the screening assay, highly suspect fume condensates can be further tested using appropriate experiment animal models. Experimental designs using intraperitoneal (Ref. 73) or intratracheal administration are relatively economical. Chronic inhalation experiments, which more closely mimic the type of exposure potentially received by welders, can be very expensive.

6. Mutagenicity and in vitro carcinogenicity screening of fume condensate. A number of tests are available to screen for mutagenic and carcinogenic potential (Ref. 76). The most well known of these is the Ames test of mutagenicity in *Salmonella.* This test has gained some acceptance because of the good correlation between results in it and *in vivo* carcinogenicity (Ref. 77). However, most results are on organic compounds. Therefore, any test proposed must be validated for solid inorganic compounds.

In the Ames test, samples of stainless steel welding fumes have been found to be mutagenic, whereas fumes from mild steel welding have not (Ref. 78). Based on these results, a study on stainless steel fume in rats has been initiated (Ref. 78). Therefore, process changes that minimize mutagenic potential should be examined for welding stainless steel. After each change, fumes should be collected and bioassayed. Other chromium- and nickel-containing fumes should be examined.

Other *in vitro* assays that measure mutagenic and carcinogenic potential (such as sister chromatid exchange (SCE) or *in vitro* transformation) should also be considered (Ref. 76). Positive screening results in any individual assay or battery of assays should be verified in appropriate animal studies.

Chapter 1

The Exposure

Details of the physical environment created by the welding process are presented in Appendix A. They are summarized here to provide the reader with the necessary background to evaluate information presented in Chapters 2, 3, and 4.

Welders are potentially exposed to the particulates, gases, radiation, and noise created as byproducts of most of the welding processes currently in use. The hazard potential of this exposure will depend primarily upon the composition and quantity of the fumes and gases generated and the spectrum and intensity of radiation emitted, which in turn depends upon the welding process employed and the metal being welded. It also depends upon the length of time spent in this type of environment and the effectiveness of control measures, such as ventilation and personal protective equipment.

Over 80 different welding processes in commercial use have been identified by the American Welding Society (Ref. 79). The health literature concerns exposure to some of the most widely used processes, including shielded metal arc welding (SMAW), also called covered electrode welding; flux cored arc welding (FCAW); gas metal arc welding (GMAW); gas tungsten arc welding (GTAW); and oxygas welding. All of these produce fumes, gases, radiation, and noise to varying degrees.

Fumes

In arc welding, fumes arise principally from the filler metal of consumable electrodes and any covering or core materials they may contain. The composition of the filler is usually similar to that of the metal being welded. Vaporization, then incomplete reaction with atmospheric oxygen, followed by rapid condensation are the main steps in their formation. The base metal weld pool is much cooler than the electrode tip, so the base metal is a significant contributor to total fumes only when it contains a volatile protective coating or volatile alloying elements. The total fume generation rate in any welding process is affected by the welding current, arc voltage, electrode polarity, electrode diameter, speed of welding, and welding practices.

In the shielded metal arc process, the consumable electrode is coated with materials that decompose to provide a protective gas shielding to the molten weld metal and fluxing. Therefore, in addition to the electrode, metal fume components arise from coating materials, which often include cellulose, calcium and magnesium carbonates, fluorspar (CaF_2), rutile (TiO_2), silicate minerals (feldspar, mica, asbestos), clay, sodium and potassium silicates, iron, silicon, and manganese. Fume composition varies widely depending upon the coating compositions, which may be divided into four classes: acidic, rutile, neutral, and basic (containing CaF_2). Tables A3 and A4 in Appendix A present the major components formed when steel is welded with electrodes having each of these types of coating.

This process is widely used for welding carbon and low alloy steels in structural applications. Table A5 lists the concentration of selected elements in fumes from using electrodes for steels and nickel alloys.

Because of the added burden of the coating materials, a relatively high fume generation rate is produced with this process. The rate increases with increasing current.

In the flux cored arc process, a continuous filler wire electrode with a core containing fluxing, deoxidizing, and slagging agents is used. Some electrodes contain additional materials that decompose to provide a shielding gas, while the gas is externally supplied in others. Many of the core materials are the same as those used in the coatings of shielded metal arc welding electrodes. Table A9 presents the percentages of several elements in the fumes from several common flux cored arc electrodes. Due to the added contribution from core material and the generally high level of arc current used, this process generates fumes at a high rate.

In gas metal arc welding, fumes originate from the continuous bare wire electrode (with minor contribution from base metal) only. No core or coating is present. Thus, the composition of the fumes should be much more easily predicted. Shielding is provided by an externally supplied inert gas (usually argon), argon with a small quantity of oxygen or carbon dioxide, carbon dioxide, or helium. The composition of fumes produced when welding is done with various currently available electrodes is presented in Table A12. When carbon dioxide is used as a shielding gas, a higher rate of fume generation occurs than when argon or argon-5 percent oxygen (Refs. 80 and 81) is used. The rate also increases with increasing arc current.

The gas tungsten arc process is adapted to welding a wide range of materials: carbon, low and high alloy steels, aluminum and its alloys, magnesium and its alloys, copper and copper-nickel alloys, brasses, silver, and others. Gas tungsten arc welding is similar to gas metal arc welding except that a nonconsumable (tungsten) electrode is used. An external filler wire (not in the electric circuit) may or may not be used. When filler is used, it is relatively cool compared to the arc temperature, yielding generally a low rate of fume generation. When no filler is used (and fusion is produced by melting together the two base metals) the rate is insignificant (Ref. 81). The plasma arc welding process is similar to the gas tungsten arc process, and for this reason the fume generation rates are generally similar.

The submerged arc process is used to weld relatively thick plates at high metal deposition rates, commonly plain carbon and low alloy steels (Ref. 82). In use the arc is not visible; it is submerged under a flux. The end of the electrode and the molten weld metal are surrounded and shielded by an envelope of molten flux, upon which rests a layer of unmelted flux. Since the arc and the weld metal are protected by flux, the fume generation rate is minimal when compared to those of shielded metal, gas metal, and flux cored arc welding. The major source of fumes is fluorides in the flux (Ref. 82).

Significant levels of fumes may be generated by the oxygas process in welding galvanized steel or an alloy containing a volatile metal (Ref. 82). Otherwise, the rate of fume production is relatively low, since the temperatures produced by the oxygas flame are much lower than arc temperatures (Ref. 82).

Table A16 and Figure A1 in Appendix A emphasize the relative fume producing capabilities of various welding processes.

In addition to fume level, duration of exposure to these fumes, location of welding, use of personal protective equipment, and engineering controls, the effects of welding fumes on the health of welders will also depend upon the composition and particle size distribution of the fumes. Some fume constituents may pose more of a potential hazard than others, depending upon their inherent toxicity. Of special concern are silica or silicates or both, fluorides, copper, chromium, nickel, and manganese.

The principal source of silica or silicates or both is the coating of shielded metal arc electrodes and the core of flux cored arc electrodes, present in the form of silica, ferrosilicate, kaolin, feldspar, mica, talc, and waterglass (sodium silicate). Some low alloy steels and aluminum alloys also contain elemental silicon. The crystalline forms of this compound are responsible for producing silicosis. However, attempts to locate crystalline phases of silica in welding fumes have been unsuccessful (Refs. 70 and 71).

The major source of fluorides in welding fumes is also the covering on shielded metal arc electrodes and the core of flux cored arc electrodes. Basic (low hydrogen) covered electrodes and self-shielded flux cored electrodes contain large amounts of fluorspar (calcium fluoride). The flux used in submerged arc welding is another source. The fluoride compounds present in the fumes from basic covered electrodes have been found to constitute 5 to 30 percent (as fluorine) of the total fumes. These are present mainly as calcium, sodium, and potassium fluorides.

The highest concentrations of copper are found when copper and its alloys are welded. Another minor source is from copper-coated gas metal arc electrodes.

Chromium arises in the fumes when stainless and high alloy steels are welded. The hexavalent, trivalent, and zero valent oxidation states of this element are present in the fumes. Some compounds of hexavalent chromium have been identified as carcinogens and mutagens (Refs. 66 and 83). Stern found that the total concentration of chromium in the fumes from welding mild steel (unalloyed) was less than 0.05 percent in the shielded metal arc process and 0.005 percent in the gas metal arc process. Shielded metal arc welding of stainless steel (15 to 25 percent Cr) with basic or rutile type covered electrodes yielded a fume content of 2.4 to 6.4 percent chromium, while it ranged from 9.8 to 13.8 percent chromium with the gas metal arc process. Of the chromium in the stainless steel welding fumes, Stern (Ref. 78) found 5 to 33 percent to be in the zero or trivalent state (all of which was insoluble in water), and 67 to 95 percent to be in the hexavalent oxidation state (of which 0 to 13 percent was insoluble in water), when the shielded metal arc process was used. In gas metal arc welding of stainless steel using argon or an argon-oxygen mixture, 98 to 99.86 percent of the chromium in the fume was zero or trivalent (all of which was water insoluble), and 0.14 to 2 percent was hexavalent (of which 60 to 90 percent was insoluble in water) (Ref. 78). The relative amount of hexavalent chromium in the fumes from the gas metal arc process depended upon the amount of oxygen in the shielding gas (Ref. 78). In contrast, Virtamo and Tuomola (Ref. 84) found that in using basic covered electrodes to weld stainless steel (18 percent Cr), 1.8 to 3.1 percent of the fumes was chromium, of which 0.57 to 2.2 percent was hexavalent. Fumes from a rutile type covered electrode were 3.6 percent chromium, of which 2.5 percent was hexavalent. The gas metal arc process produced fumes of 0.40 to 0.96 percent chromium, 0.2 to 0.32 percent being hexavalent (Ref. 84).

Nickel is produced in fumes produced in welding stainless steel (typically 8 to 15 percent Ni) and nickel alloys.

Manganese is used in the coating of some shielded metal arc electrodes and in the core of flux cored arc electrodes. In a recent study of Pattee et al. (Ref. 70), manganese accounted for 0.3 to 8.8 percent and 1.0 to 13.5 percent of the total fumes produced by these two processes, respectively. Also, special steels with a high manganese content are another source of manganese oxides in welding fumes (Ref. 80).

Although fumes generally arise from the electrodes and the base metals, other sources may contribute a considerable fraction to the fumes in special cases. Zinc from galvanized steel and pigments and decomposition products from welding on painted surfaces are of special importance.

The particle size distribution is an important factor in determining the hazard potential of welding fumes, since it is an indication of the depth to which particles may penetrate into the respiratory system and the percentage of particles that will be retained therein. Particles in the range of 1 to 7 μm in diameter represent the most serious hazard due to penetration into the alveolar region (Ref. 85). Welding fume particles are essentially all less than 1.0 μm in diameter (Refs. 70, 86, and 87).

Hedenstedt et al. (Ref. 88) found that about 90 percent of the fume particles generated during the welding of stainless steel were less than 1 μm in diameter. However, the particles produced by the shielded metal arc process were relatively larger than those formed during gas metal arc welding. Stern (Ref. 78) found that the mass median diameter produced in shielded metal arc welding of stainless steel was approximately 0.3 to 0.5 μm, whereas with the gas metal arc process it was approximately 0.25 μm. By extrapolating from a least squares plot, Heile and Hill (Ref. 81) estimated the mean particle diameters to be 0.03 μm and 0.12 μm from welding steel by the gas metal arc and flux cored arc processes, respectively. During gas metal arc welding with carbon dioxide, the mean particle size was greater than when argon was used. Although most welding fume particles are less than 1 μm in diameter when formed, results of one study (Ref. 85) indicate that they may agglomerate and increase in size with time.

Gases

Ozone, nitrogen oxides, carbon monoxide, and carbon dioxide are the principal gases produced during arc and oxygas welding. Phosgene, hydrogen chloride, and diacetyl chloride are produced incidentally from the photochemical (welding radiation) oxidation of chlorinated hydrocarbon solvents that may be present in the atmosphere.

Ozone is produced from atmospheric oxygen in a photochemical reaction by radiation of a wavelength shorter than 210 nm emitted from the welding arc. At wavelengths shorter than 175 nm, the process is so effective that virtually none of this radiation penetrates further than a few centimeters beyond the arc; the effectiveness decreases with increasing wavelength (Ref. 89). At wavelengths in the range of 220 to 290 nm, ozone absorbs ultraviolet energy and decomposes back to diatomic oxygen (Ref. 28).

The rate of formation of ozone depends upon the wavelengths and the intensity of ultraviolet radiation generated in the arc, which in turn is affected by the material being welded, the type of electrode being used, shielding gas (if it is used), and the welding variables (such as voltage, current, and arc length) (Ref. 90).

Gas shielded arc welding processes present a much greater problem from ozone production than the shielded metal arc (Refs. 28, 91, and 92) or flux cored arc (Ref. 28) processes. During the welding of aluminum and aluminum-magnesium alloys, Vorontsova (Ref. 91) found no detectable amounts of ozone when the shielded metal arc process was used; when argon was used as the shielding gas, the gas metal arc process produced ozone levels approximately four times higher than did gas tungsten arc welding. Sampling 6 inches from various arcs, Lunau (Ref. 28) found low levels (0.12 to 0.24 ppm) when mild steel was welded with either the shielded metal arc or flux cored arc process. Higher levels (0.27 to 2.1 ppm) were found during gas tungsten arc welding of mild steel. The highest (2.3 to 14.5 ppm) were attained during argon shielded metal arc welding of aluminum or aluminum-based alloys (Ref. 28). One possible explanation for the higher production of ozone during gas metal arc welding, relative to the shielded metal arc or flux cored arc processes, is the higher level of fumes produced by the latter, which tends to block the emission of ultraviolet radiation.

The formation of ozone during submerged arc welding should be negligible, since there is no visible evidence of an arc. During oxygas welding, the flame is not hot enough to emit radiation of sufficient energy to generate ozone.

Changes in shielding gases and metals produce their effect on ozone generation rate by causing arc spectral changes. Those that increase the intensity of radiation at wavelengths shorter than 210 nm will increase the ozone generation rate. It appears that the highest levels of ozone are generated by the use of argon shielding in gas metal arc welding of aluminum (Refs. 28 and 89). Frant (Ref. 89) claimed that ozone was generated at 15 to 20 times higher levels when argon was substituted for helium in the welding of aluminum or copper. Similarly, less ozone is generated when carbon dioxide, rather than argon, is used (Ref. 89).

Alloying elements can play an important role in the ultraviolet light spectrum and, consequently, in ozone production. Lunau found that in welding aluminum the addition of 5 percent magnesium to the aluminum partially suppresses ozone generation, whereas the addition of 5 percent silicon enhances it.

Oxides of nitrogen are formed by the direct oxidation of atmospheric nitrogen at the high temperatures produced by the arc or flame (Refs. 93, 94, and 95). Table A21 in Appendix A presents the levels of nitrogen dioxide found during various types of welding operations; in only one case did the concentration in or around the welder's helmet appreciably exceed 2 ppm.

Carbon dioxide and carbon monoxide are formed by the decomposition of organic compounds in electrode coatings and cores, from inorganic carbonates in coatings, from carbon in weld metal, and in the oxyacetylene flame. Carbon monoxide is generated by the decomposition of carbon dioxide used in gas shielded arc welding processes.

Radiation

Electromagnetic radiation from the ultraviolet, visible, and infrared portions of the spectrum is emitted by most arc welding processes. The arc generates line spectra characteristic of the materials involved, superimposed upon a continuum of radiation (Ref. 39).

Virtually no radiation of a wavelength shorter than around 175 nm should strike the welder; as previously stated, absorption of this by oxygen is so effective that the intensity should be reduced essentially to zero after it passes through a few centimeters of air (Refs. 28 and 89). Atmospheric oxygen will also absorb the 175 to 210 nm wavelength radiation.

The intensity of at least the radiation in the ultraviolet portion of the spectrum is also attenuated by the fumes produced. During shielded metal arc welding, UV irradiances at 4.8 m from the electrode were found to be 12 to 100 times higher when a high rate (not specified) blower was used to remove fumes than when only natural ventilation was available (Ref. 96). UV irradiance increases with increasing current, and Lyon et al. (Ref. 97) demonstrated that the increase in actinic ultraviolet (200 to 315 nm) irradiance was roughly proportional to the square of the current for gas tungsten arc, gas metal arc (CO_2), and flux cored arc (CO_2) welding of mild steel. Reflection from highly polished surfaces also increases exposure to radiation.

Noise

All of the commonly used welding processes produce noise. Levels for various processes measured with a type 1 meter by Rodman et al. (Ref. 98) are presented in Table A23. In addition to noise produced by the welding equipment itself, welders are often exposed to noise from other operations in their workplace.

Other Factors

Heat may present a problem when welding is done in confined spaces or when metal that has been preheated to improve welding characteristics is used. In addition to any direct effects, heat and exertional stress may also increase a welder's susceptibility to the effects of other harmful agents.

Chapter 2

Effects of Welding on Health

A major problem in evaluating many of the human toxicity studies has been the lack of information provided by the author(s) as to the nature of exposure: namely, the welding fume and gas concentrations in welder's breathing zones, the type of process and protection being used, length of time spent working in confined spaces, and other pertinent data. Another consideration is the way in which a particular group of welders was chosen for evaluation. A random sample is essential. If only healthy workers are examined, for example, the results would be different than if the study included welders presently on sick leave or unemployed. Because of the influence of cigarette smoking, alcohol use, drug use, diet, previous occupational exposure, and physical constitution, among other factors, upon the health of a population, the effects of welding exposures cannot be isolated and studied without considering these variables. For these reasons, there are not very many conclusive reports on the health of welders in the available literature. It is in this light that reports on the effects of welding on health are presented.

Respiratory Tract

The discussion is divided into sections dealing with acute and chronic lung conditions, although there is some overlap. Metal fume fever is considered in a separate chapter, even though lung damage is presumed to play a role in this condition.

Acute Diseases Due Directly to Occupational Exposure

In examining acute pulmonary diseases in welders, it is always necessary to evaluate the adequacy of any respiratory protective equipment and ventilation in use. Acute pulmonary diseases in welders working in inadequately ventilated confined areas and exposed to toxic concentrations of mixtures of various welding fumes and gases are described in case reports dating from 1929 to the present. No specific intoxicant could be identified in some reports (Refs. 99, 100, and 101). The development of varying degrees of pulmonary edema, with the possibility of heart failure, acute pneumonia, bronchitis, or combinations thereof, is often observed, regardless of the toxic substance(s) involved.

The patient may become cyanotic, developing severe breathing difficulty and chest pain, prior to becoming unconscious. Death can occur in minutes or hours. If the patient recovers, an acute pneumonia is likely to persist for up to two weeks.

Inhalation of sufficient concentrations of nitrogen oxides (Ref. 41), ozone (Ref. 29), phosgene (Ref. 102), or fumes containing cadmium (Ref. 103) can cause acute pulmonary edema, acute pneumonia, and acute bronchitis. The more important hazardous agents are discussed individually.

Ozone

In certain welding situations, exposure to ozone is a serious potential health hazard. Since ozone is generated by the action of ultraviolet radiation on molecular oxygen, it can be present in substantial concentrations several feet from the arc. It is a special hazard in GMA and GTA welding. Argon shielded welding produces more ozone than helium shielded arcs (Ref. 104). Appendix A further discusses the formation of ozone.

Ozone is an irritant gas. At concentrations above 0.1 ppm, drying of the mucous membranes of the mouth, nose, and throat occur. At approximately 1 ppm, headache and general irritation are noted. Visual disturbances have resulted after exposure to ozone levels of 0.2 to 0.5 ppm for 3 hours. Respiratory tract irritation, pulmonary congestion, and edema may result from exposure to 1.5 to 2 ppm inhaled for periods of 1 hour or more. Deterioration of defensive mechanisms of the lung has been reported (Ref. 29). Exposure to 1 ppm of ozone for 10 minutes can precipitate asthma attacks. Respiratory irritation during mild exercise may occur with exposure to 0.5 ppm for 30 to 60 minutes. Both effects are reversible (Ref. 29).

Kleinfeld and Giel, 1956 (Ref. 105), studied severe acute ozone intoxication in welders. Pulmonary edema and hemorrhage, possibly fatal, or temporary pulmonary insufficiency in less severe exposures are characteristic. Secondary effects include irritation of the mucous membranes, headache, and lethargy. Low concentrations produce general depression or sleep in certain individuals.

Three features of ozone intoxication are noteworthy. The first is the delayed onset and severity of the condition. The second is the fact that symptoms such as marked dyspnea, chest pain, or cough are much worse than the clinical signs, which indicate only minimal pulmonary involvement. The third is that residual symptoms persist for months after exposure, even though the physical signs and chest x-ray films indicate the lungs have cleared (Ref. 105).

Moles and Collins, 1957 (Ref. 106), studied less severe effects of ozone exposure in GMA welders over a 3 month period. Ozone levels in the workroom air exceeded 0.1 ppm in all determinations. Eye and respiratory tract irritation complaints were voiced when the inert gas used was a mixture of 98 percent argon and 2 percent oxygen, but not when pure argon, helium, or carbon dioxide was used. Increased current density and gas flow rate increased the severity of these complaints. Other workers at various distances and directions from the arc also simultaneously experienced eye and respiratory tract complaints due to ozone formation by the ultraviolet radiation. Symptoms disappeared after termination of welding with the $Ar\text{-}O_2$ mixture, leaving no evidence of residual effects or disability. Use of respiratory protective equipment and ventilation (if any) was not specified.

A case of acute pulmonary edema in a 47-year-old welder developed 4.5 to 5 hours following his welding in a 5,000 gallon aluminum tank with inadequate ventilation and no respirator. The metal had been preheated and was then welded by the argon tungsten arc process. Recovery was rapid; lung function values returned to normal within 11 days of the incident. Kurta, 1976 (Ref. 107), noted that the toxic agents inhaled were most likely a mixture of ozone and nitrogen oxides, although no measurements of these gases were reported.

Kleinfeld, 1970 (Ref. 6), reported the case of a 51-year-old welder who developed ozone pneumonitis after 240 minutes of exposure to gas shielded arc welding (GMAW) in a poorly ventilated area with an ozone level of 1.8 mg/m^3 (0.9 ppm). NO_2 levels were stated to be minimal. The lungs cleared in 6 days.

Three other cases of ozone intoxication in male welders were reported by Kleinfeld et al., 1957 (Ref. 108). These men worked in a 200 x 90 x 10 ft room with the welding area in one corner. No supplementary ventilation was provided. One GMA and two GTA welding machines were used in this area by 8 workmen on two shifts. The work was done on nickel. Atmospheric ozone levels reached 9.2 ppm, but levels of nickel carbonyl and nitrogen oxides were stated to be negligible.

The first case, in a 51-year-old, was marked by a sudden severe headache, substernal pressure, and dyspnea, all of which occurred during GMA welding. Pulmonary edema developed. A chest x-ray film showed diffuse peribronchial infiltrates similar to acute pneumonia. After remaining in critical condition for two days due to persistent lung congestion, the welder slowly recovered over the course of two weeks, when chest x-ray films showed the lungs had cleared. Nine months later, he still experienced unusual fatigue and exertional dyspnea. The man had been a "conventional" arc welder for 10 years, but had switched to operating the GMA welding machine two weeks before the episode of pulmonary edema.

Another case occurred in a welder operating a GMA welding machine for three days; previously, he had 14 years of experience in arc welding. He noted burning of the eyes and throat for 3 days, a cough, a sensation of choking, and dyspnea. A chest x-ray showed small scattered infiltrations in both lung fields. He was hospitalized and treated for bronchopneumonia. Nine days later he was discharged with cleared chest x-ray films. For the next 9 months, he still experienced fatigue and frequent head colds, although clinical examinations revealed no abnormalities (Ref. 108).

Nitrogen Oxides

Overexposure to nitrogen oxides can cause pulmonary edema, acute pneumonia, and bronchitis and can lead to chronic lung diseases including emphysema and bronchiolitis fibrosa obliterans (Ref. 109). Nitric oxide (NO), nitrogen dioxide (NO_2), dinitrogen trioxide (N_2O_3), dinitrogen tetroxide (N_2O_4), and dinitrogen pentoxide (N_2O_5) generally occur together. N_2O_5 is unstable above 0° C but may occur in the presence of ozone. When inhaled, certain nitrogen oxides may react with water to yield nitrite and nitrate ions. Nitrosation by N_2O_4 (and possibly NO), changes in acidity with the formation of nitric acid, oxidations by N_2O_3 and N_2O_4, nitrite ion effects, and the reaction of NO with hemoglobin may be responsible for the toxicity (Ref. 110).

Exposure to nitrogen oxides can lead to violent dry coughing; high concentrations cause laryngospasm. Pulmonary edema and methemoglobinemia lead to cyanosis, with a possibility of death in a few minutes. Recovery usually follows hospitalization of the subject. The pulmonary edema may lead to a chemical pneumonitis, which then can induce a chronic inflammatory response leading to fibrotic changes. A predisposition to pneumonia during recovery has been suggested (Ref. 110).

Cadmium

Cadmium fume exposure is rare (Ref. 111). Hazard occurs whenever cadmium alloys are welded in an enclosed space (Ref. 112). Cadmium-containing or cadmium-plated metals, including some stainless steels, will evolve the oxide fumes when heated during welding processes. Cadmium fume overexposure represents an extreme health hazard and medical emergency (Ref. 113).

A number of case histories of cadmium overexposure in welders have been reported (Refs. 114 through 117). Intoxication may develop immediately or have a delayed onset. The clinical picture may at first resemble metal fume fever (Ref. 118), but is much more severe; it includes mild throat irritation and a foul taste in the mouth initially, followed in one to two hours by acute upper respiratory tract symptoms of cough, dyspnea, and pain, along with headache and chills. This progresses to severe pulmonary edema and possibly bronchopneumonia. Acute gastroenteritis may also occur. In cases of metal fume fever, in contrast, the symptoms usually subside in 12 hours or less (Ref. 114).

According to Christensen and Olson, 1957 (Ref. 115), the chest x-ray picture of acute cadmium intoxication resembles bronchopneumonia, but these findings are out of proportion to the clinical presentation. Whether or not permanent lung or kidney injuries result in nonfatal cases remains to be investigated, according to Beton et al., 1966 (Ref. 114). Christensen and Olson emphasized that cadmium fumes are not exceedingly irritating and do not have a pronounced odor. They estimated 2500 mg/m^3 of cadmium fumes in air is lethal to man (Ref. 115).

General Acute Respiratory Diseases in Welders

A higher incidence of pneumonia or other acute respiratory symptoms in welders would be a possible indication of harmful effects of welding fumes and gases. In reviewing the London Decennial report for 1951, Doig and Challen found that deaths from all causes in welders were slightly higher than expected, 1196 compared to 1092 expected, yielding a standard mortality ratio of 110 (observed-to-expected mortality; if observed and expected are the same, the value would be 100). A substantial part of the excess was due to pneumonia; there were 70 deaths from pneumonia in welders against 31 expected. This increased pneumonia risk was not age-related, but seemed to be constant throughout the welders' working lives. It was unclear whether the acute pneumonias were caused by infectious microorganisms or by occupational overexposures to toxic substances in the welding environment (Ref. 94).

In other welders, Collen, in 1947 (Ref. 119), came to an opposite conclusion regarding pneumococcal pneumonia rates. No significant difference between 12,100 shipyard welders and all other shipyard workers exclusive of welders was found for either morbidity or mortality frequency due to pneumococcal pneumonia. The case fatality rates were 5.2 percent for welders and 5.8 percent for all other workers. The pneumonia in the welders, who were treated at the Permanente Foundation Hospital, were no different in severity, incidence of complications, or treatment days required than for all shipyard workers (Ref. 119).

In a cross-sectional study undertaken in Finland by Antti-Poika et al., 1977 (Ref. 7), 157 arc welders were compared to 108 male controls having similar smoking habits and socioeconomic status. Welders were exposed to fumes from basic electrodes used for mild unpainted steel for at least 3 years, at least 3 hours per day in shops; 25 welders had worked in enclosed spaces. Ventilation or respiratory protection in use was not specified. Previous exposures to dust in other occupations occurred in 59 welders (38 percent) and 17 controls (16 percent). Questionnaires on acute respiratory disease experience revealed the prevalences shown in Table 1. Only colds, hoarseness, and sore throats were significantly more prevalent in the welders than in the controls. The authors were able to conclude that serious acute disorders of the respiratory tract were no more common in male welders than in matched controls.

Siderosis and Mixed Dust Pneumoconiosis

Protracted inhalation of welding fume particles leads to accumulation in the lungs of otherwise healthy workers (Ref. 120). Deposits of the welding particles and dusts in the lungs can be noted on chest x-ray films as areas that are denser than the normal appearing lung fields. These densities may be small nodules, termed pinhead or micronodular opacities, or linear markings. In some chest x-ray films of welders, a combination of the two, that is, a reticulonodular pattern, is seen. Some authors refer to this as "snowstorm" lungs. This is not necessarily a clinical lung disease, although chest x-ray findings sometimes can be mistaken for asbestosis (Ref. 58). A number of cases in which biopsy or autopsy have been performed have been reported (Refs. 58 and 121 through 129).

The prevalence of chest x-ray evidence of mixed-dust pneumoconiosis in welders varies from none at all (Ref. 8) to percentages as high as 71 percent of welders examined (Ref. 130). The details of many epidemiologic investigations are presented below and are summarized in Table 7 at the end of this section. In several of the

Table 1
Prevalence of acute respiratory diseases in welders in Finland

Condition	Welders			Controls			P
	Number questioned	Positive response number	Positive response %	Number questioned	Positive response number	Positive response %	
More than 2 colds per year	115	27	23	82	9	11	<0.05
Frequent attacks of nasal catarrh without cold	157	22	14	98	10	9	N.S.
Frequent attacks of hoarseness without cold	157	33	21	81	5	6	<0.01
Frequent attacks of sore throat without cold	157	68	43	81	22	27	<0.05
Pneumonia	156	14	9	106	8	8	N.S.
Acute bronchitis	157	21	13	106	10	9	N.S.

N.S. = Not statistically significant

Antti-Poika et al., 1977 (Ref. 7).

more recent studies, no abnormal roentgenological changes were observed, possibly indicating improved conditions. In some studies, neither the type of welding process nor the adequacy of industrial hygiene measures (if any were provided) were presented in detail by the authors.

In groups in which exposure to welding fumes has resulted in abnormal lung film changes, the frequency and severity of those changes have been shown to be related to the duration of occupational exposure. Dreessen et al., 1947 (Ref. 3), studied arc welders in steel ship construction at 7 shipyards in the United States. From a random sample that included over 3,000 welders (about 15 percent of the total welding workforce), 4563 x-ray films were evaluated. More men than women had exaggerated lung field markings: 36.1 percent and 17.3 percent, respectively. Male welders and nonwelders were compared on the basis of age groups to determine the presence of abnormal chest x-ray markings. The results are shown in Table 2. Age was proportional to the prevalence of exaggerated lung field markings in both welders and nonwelders working in the shipyards.

Dreessen et al. also evaluated the association between the length of occupational exposure to welding and the occurrence of exaggerated lung field markings. They divided the sample into those men with and those without previous industrial exposures, obtaining the results shown in Table 3.

Table 2
Prevalence of abnormal chest x-ray films in U.S. shipyard workers by age

Age (yr)	Male welders, %	Male nonwelders, %
15-24	2.6	1.4
25-34	8.4	5.3
35-44	9.9	6.3
>45	9.9	11.5

Dreessen et al., 1947 (Ref. 3).

Table 3
Prevalence of abnormal chest x-ray films in U.S. shipyard welders by duration of exposure

Duration of welding exposure	Male welders with no previous industrial exposure, %	Male welders with previous industrial exposure, %
<10 months	5.3	5.9
10-19	5.4	7.6
20-29	6.2	10.6
>30	11.0	12.3

Dreesson et al., 1947 (Ref. 3).

Table 3 shows that the duration of exposure was associated with an increased prevalence of exaggerated lung field markings, and it is possible that prior exposure to other dusty industrial environments could have influenced the abnormal chest x-ray films as well. Further analysis of the chest films suggested welders' siderosis in 1.9 percent of the welders examined, but none in the nonwelders. Among white males, the prevalence was 3.4 percent; for nonwhite males it was 1.3 percent; and for white females the prevalence was zero.

A study by Britton and Walsh, 1940 (Ref. 131), reviewed chest x-ray films of 286 welders with 5 years or more (average 9 years) of exposure to arc or oxygas welding. Their ages ranged from 22 to 63 years. Most welders worked in well ventilated areas. There were 121 arc, 49 oxygas, and 66 "mixed" welders, all working in 14 plants involved in machinery manufacture. Mottled or stippled lung fields were revealed on chest x-ray films of 24 (8.4 percent) of the welders. Increased hilar or linear lung markings were found in 47 (16.4 percent). In 186 welders (65.0 percent), the chest x-rays were considered normal.

Garnuszewski and Dobrzynski, 1964 (Ref. 4), reported a chest x-ray survey of 307 shipyard welders. Abnormalities appeared in chest films of 192 (60 percent) of the welders and were interpreted as pulmonary siderosis coexisting with silicosis. In repeat chest films taken 1-1/2 years later, 31.7 percent of the cases showed a rapidly progressive course. The prevalence of this abnormality was directly proportional to the duration of welding exposure. All workers with more than 10 years' experience were affected. Inadequate ventilation in ship compartments and dockyard shops was probably one factor influencing the x-ray abnormality.

Dobrzynski, 1973 (Ref. 5), reviewed chest x-ray films of 1027 manual arc welders during the period from 1961 to 1972, with respect to type of exposure (see Table 4). The duration of exposure to welding was associated with the frequency of appearance of abnormal chest films. Only 1 percent of group I welders (hull welders in confined space) with abnormal chest films had worked for less than 3 years, while over 60 percent of the cases occurred in welders with over 13 years of exposure. The chest x-ray picture was characterized by linear and micronodular shadows, which the authors classified as pneumoconiosis due to exposure to iron, silica, and other dusts in the hull welders' environment. Dust levels were not reported.

Of 210 arc welders from various industries in Santiago, Chile, 32 had clear-cut pulmonary siderosis, according to Schuler et al., 1962 (Ref. 58). The authors claimed that the welders were exposed solely to iron oxide fumes, without external silica contamination. However, electrodes were covered with carbonates and oxides or a fine mesh of asbestos. Ages of the welders ranged from 30 to 60; the mean was 45 years. The majority had been welding for at least 15 years, with a range of 5 to 25 years. Normal lung function test results and the absence of respiratory disease symptoms were noted in 84 percent of the welders, including those with the abnormal chest films.

Kierst et al., 1964 (Ref. 132), reported chest x-ray

Table 4
Prevalences of abnormal chest x-ray films in manual arc welders according to working conditions

				Mean age (years)		Mean exposure (years)	
Group	No.	Working conditions	Abnormal chest x-ray	Welders with abnormal chest x-rays	Welders with normal chest x-rays	Welders with abnormal chest x-rays	Welders with normal chest x-rays
I	650	Hull welders in confined space	77 (12%)	38.6 ± 7	32.9 ± 9	14.0 ± 5	9.2 ± 6
II	46	Container welders outside shipbuilding industry	6 (13%)	41.5 ± 6	40.3 ± 8	13.4 ± 5	11.5 ± 9
III	156	Shipyard welders in open air	5 (3%)	35.5 ± 5	34.6 ± 8	11.0 ± 2	10.2 ± 5
IV	175	Welders in other industries	4 (2%)	37.7 ± 9	38.2 ± 9	11.7 ± 7	12.2 ± 6

Dobrzynski, 1973 (Ref. 5).

findings for a group of 171 welders employed in the Gdynia Shipyards in Poland, who were outpatients registered at an occupational disease medical facility. Exposure to fumes from rutile covered electrodes used in SMA welding was noted. No levels of fumes or other information were provided. The patients and hull fitters were evaluated for pulmonary disease and lung function in groupings based upon chest x-ray findings. The results are presented in Table 5. The ages of the welders ranged from 18 to 62 years; only 12 were over age 45. The length of welder employment was 1 to 32 years. Chest x-ray films were normal in 30 (19.3 percent, 2 to 16 years exposure). In 63 (36.9 percent, 2 to 32 years exposure), there were localized shadows considered to be on the borderline of normal. There were 64 welders (37.4 percent, 2 to 21 years exposure) with scattered nodular shadows, and 11 welders (6.4 percent, 6 to 16 years exposure) with "snowstorm" type shadows (diffuse micronodular pattern). Both the nodular and micronodular chest films were classified as mixed siderosis and silicosis. Among a control group of 43 hull fitters with 1 to 15 years of occupational exposure in the shipyards and aged 18 to 45 years, borderline chest x-ray findings were noted in 18 (41 percent); the rest had normal chest films. For both welders and controls, the percentage of abnormal chest films was proportional to the duration of occupational exposure.

The relatively mild clinical course of the disease was emphasized. The siderosilicosis caused little pulmonary fibrosis or interference with pulmonary circulation, even after long exposures to welding fumes, in contrast to true silicosis, which would have caused obvious disability.

Spacilova and Koval, 1975 (Ref. 129), compared chest films of arc welders working in confined areas to those of arc welders working in well-ventilated conditions in industries in Prague. The mean age, the mean duration of occupational exposure, and the number of welders for both groups are listed in Table 6. In 18 (78 percent) Group A workers and 12 (96 percent) in Group B, chest films revealed abnormalities. In 26 of the 30 films, alterations were slight; but, in 4 chest films of welders in Group A with 25 or more years of occupational exposure to welding, nodular opacities indicative of fibrosis were noted. The substance(s) causing the opacities was not identified. Eight of those from both groups had chronic bronchitis and reduced pulmonary function. Pulmonary function was normal in the remainder.

Gillon and Marchand, 1963 (Ref. 133), found 27 cases (5.1 percent) of reticular or nodular densities in both lung fields, particularly in the 7th and 8th intercostal spaces, without apical involvement or diaphragmatic or cardiac abnormalities, after examining chest x-ray films of 500 arc welders. All welders had more than 5 years of occupational exposure to SMA or GTA welding. Those with silica exposure or histories of lung disease were excluded from the study. Lung function tests (not specified) remained normal in the welders with abnormal chest films.

There appears to be no reliable correlation between the prevalence of pneumoconiosis and frequency of appearance of lung function deficit (Refs. 7, 8, and 134 through 136). Studies that have investigated both are presented in the following section.

Table 5
Clinical, functional, and radiological respiratory system abnormalities in Polish shipyard welders

Chest x-ray picture		Number (%) of welders	Number (%) of controls	Clinical and functional findings in welders
Group I:	no chest x-ray abnormality	30 (19%)	25 (59%)	None
Group II:	slight localized shadowing	63 (37%)	18 (41%)	None
Group III:	linear shadows and small nodular opacities in peripheral lung fields	64 (37.4%)	-	Chronic bronchitis, pulmonary emphysema, and effort dyspnea in 6 welders; decreased vital capacity in 19 welders
Group IV:	"snowstorm appearance," diffuse micronodular opacities	11 (6.4%)	-	Chronic bronchitis and effort dyspnea in 11 welders; beginnings of corpulmonale in 5 welders; decreased vital capacity in 5 welders

Kierst et al., 1964 (Ref. 132).

Table 6
X-ray abnormalities and working conditions of welders in Prague

	No.	Abnormal x-ray	Age: Mean age	Age: Range	Exposure (yrs): Mean duration	Exposure (yrs): Range	Working conditions
Group A	23	18	42	30-56	18	10-32	13% of yearly worktime in confined spaces; dust levels of 49-145mg/m^3
Group B	14	12	46	31-64	16	3-34	Open-air welding; no dust; basic electrodes used

Spacilova and Koval, 1975 (Ref. 129).

Table 7
Prevalences of welders with chest x-ray pictures characteristic of siderosis or mixed-dust pneumoconiosis

Ref.	Year	Workplace	Prevalence, %	Duration of work (years)
7	1977	Nonshipyard	0	36 (mean)
8	1973	Shipyard	0	-
5	1973	Ship's hull, confined space;	12	9.8 (mean)
		container welding (nonshipbuilding);	13	11.7
		shipyard, open air;	3	10.2
		other industry	2	12.2
10	1969	Manufacturing	32	-
9	1964	Shipbuilding	34	10-35
4	1964	Shipyard	60	-
133	1963		5.1	-
58	1962	Various	15	5-25
3	1947	Shipyard	Men 36.1	-
			Women 17.3	
130	1944	Manufacturing	71	>6
131	1940	Manufacturing	35	>5

Chronic Respiratory Conditions

The chest x-ray findings, presented in the previous section, do indicate evidence of exposure of the lungs to welding fumes. A number of cases (Refs. 134 through 138) of chronic dysfunctional respiratory conditions, including pulmonary fibrosis, emphysema, and chronic bronchitis, have been reported in welders; but the association between these conditions and welding fumes and gases must be elucidated through epidemiologic investigation. An understanding of these diseases and of the pulmonary function tests used in their determination is necessary to an understanding of the discussion that follows, and they are, therefore, discussed in detail in Appendix C.

The welders' risk of developing chronic lung or respiratory tract disease depends upon many interrelated factors: welding related factors such as the availability, adequacy, and use of ventilation, respirators, and related respiratory protective equipment; fume levels and composition; and concentration of gases in the welding environment, which will differ in the confines of small containers, enclosed and open-air welding in shipyards, small shops, and large, well-ventilated industrial workplaces. Other risk factors not related to welding include aging, lack of physical fitness, cigarette smoking, urban air pollution, hereditary predisposition, and prevalence of certain lung diseases in populations living in a given geographical location.

Epidemiologic studies may contain certain biases that can distort their conclusions. Sources of bias include: the selection of study population (especially, use of the cross-sectional survey, which does not consider absent, ill, dead, or otherwise missing welders); the design of interviews and questionnaires; and the selection of a control group for any comparisons that need to be made.

A NIOSH survey of death records of males over age 20 in Washington state during the years of 1950-1971 revealed a statistically significant excess of deaths in the 1376 welders and flame cutters from all diseases of the respiratory system, including cancer, chronic bronchitis, and pulmonary emphysema ($p < 0.05$) compared to all male deaths in Washington during these years (Ref. 30).

Ten cross-sectional epidemiological studies have been reported that can contribute to the assessment of chronic lung effects in welders. These are presented in detail below. The studies are summarized in Table 25 at the end of this section.

Hunnicutt et al., 1964 (Ref. 9), compared the pulmonary functions of arc welders, under age 60 with 10 years or more of experience, to those of nonwelders, all employed at a shipbuilding plant with approximately 20,000 employees in 1962. After excluding those with significant chest and lung lesions, asthma, a past history of exposure to known irritating dusts or fumes, or with cardiovascular disease, 100 welders and 100 controls were randomly chosen. No data on welding processes or materials, ventilation, or fume and gas levels were presented. Reports of respiratory symptoms were tabulated according to smoking habits for both groups, as shown in Table 8.

Of the welders, 34 percent had x-ray evidence of pulmonary siderosis, having worked as welders an average of 18.3 years; the remainder had worked as welders for an average of 18.9 years. No other x-ray abnormalities were noted.

The prevalence of respiratory symptoms in welders who smoked cigarettes was twice that of nonsmoking welders. The controls reported respiratory symptoms less frequently than welders, regardless of smoking habits.

Pulmonary function tests revealed that 46 percent of the welders and 23 percent of the controls had impaired timed spirograms: they were unable to expel 70 percent of the maximal expiratory volume in one second ($FEV_{1.0}$). The welders also had significantly lower ($p = 0.01$) maximal expiratory flow rates and maximal midexpiratory flow volumes than did the controls. The respiratory impairment of the workers was indicative of obstructive lung disease in 37 percent of welders and in 10 percent of the controls, while restrictive lung disease was diagnosed, on the basis of pulmonary function testing, in 3 percent of welders and 11 percent of controls. Six percent of the welders and 2 percent of the controls had combined obstructive and restrictive lung disease.

When lung function was evaluated in terms of cigarette

Table 8
Prevalence of chronic respiratory symptoms in shipyard welders

Symptom	Welders (N=100)	Controls (N=100)	Smoking welders (N=71)	Nonsmoking welders (N=29)	Smoking controls (N=59)	Nonsmoking controls (N=41)
Cough	25	11	21 (30%)	4 (14%)	10 (17%)	1 (2%)
Expectoration	18	8	14 (20%)	4 (14%)	7 (12%)	1 (2%)
Wheezing	13	2	12 (17%)	1 (3%)	2 (3%)	0 (0%)
Dyspnea	12	2	9 (13%)	3 (10%)	2 (3%)	0 (0%)

Hunnicutt et al., 1964 (Ref. 9).

Table 9
Abnormal lung function prevalences in shipyard welders

	N	Obstructive lung dysfunction N	%	Restrictive lung dysfunction N	%	Combined lung dysfunction N	%	Total N	%
Welders:									
Smokers	71	29	41	2	3	5	7	36	51
Nonsmokers	29	8	28	1	3	1	3	10	34
Controls:									
Smokers	59	6	10	8	14	1	2	15	25
Nonsmokers	41	4	10	3	7	1	2	8	20

Hunnicutt et al., 1964 (Ref. 9).

smoking habits, the data shown in Table 9 were reported.

Welders who smoked had a twofold increase in prevalence of abnormal pulmonary function tests when compared to controls who smoked. Smoking and welding were stated to have a "cumulative effect" on lung dysfunction. The dysfunction, however, was considered to be subclinical, since the welders experienced no apparent respiratory difficulties.

Table 10 summarizes data on welding, smoking, pulmonary function, and symptoms of respiratory disease for the shipbuilding welders and controls studied. Hunnicutt et al. concluded that symptoms of lung disease (cough, expectoration, dyspnea, and wheezing) followed the pulmonary function results among the several groups, being present twice as often among welders who smoked as among welders who did not smoke cigarettes.

Kleinfeld et al., 1969 (Ref. 10), and Kleinfeld, 1970 (Ref. 6), compared respiratory symptoms, pulmonary function, and chest x-ray abnormalities in welders with age- and residence-matched controls with no occupational exposure to dust. The 25 welders were also divided into those with less than or more than 20 years of welding exposure (12 and 13 welders, respectively). The results are presented in Table 11.

The welders were from a plant manufacturing sheet metal products from stainless steel and, presumably, mild steel. Iron oxide fumes inside face shields ranged from 0.65 to 47 mg/m^3. Ozone concentrations were found to be negligible; 2.7 mg/m^3 of fluoride fumes were found in an air sample collected outside the face shield of an FCA welder. No other information about welding processes, materials, or type of ventilation was provided.

There were no significant differences in pulmonary function test results either between welders and controls, or between welders exposed for less than or greater than 20 years. Although 32 percent of welders had chest x-ray evidence of pulmonary siderosis, which was more frequent in the welders with over 20 years of exposure, there were no significant differences in the prevalence of cough, dyspnea, or wheezing between welders and controls. The number of cigarette smokers (at least one pack per day for 5 years or more) was similar in welders and controls. However, the small sample size should be noted.

Male welders were not at a greater risk of developing serious respiratory diseases than other males with similar smoking habits and socioeconomic status, according to Antti-Poika et al., 1977 (Ref. 7). The study included 157 currently employed arc welders who were exposed to total fume levels up to 9 mg/m^3 measured outside the helmet, using basic covered electrodes to weld, mainly, mild unpainted steels.

The welders had at least 3 years' experience and welded at least 3 hours per day in shops; 25 welders had worked in confined spaces. Fifty-nine welders (38 percent) and 17 controls (16 percent) had previous exposures to dusty occupations other than welding. Various criteria by which the welders and controls were compared are presented in Table 12.

No x-ray evidence of pneumoconiosis or any other chest findings that could be related to welding were observed.

Pulmonary function in all welders currently employed was similar to those of the controls. However, the forced vital capacity (FVC) and one-second forced expiratory volume ($FEV_{1.0}$) became significantly lower with increasing age and duration of exposure to welding fumes and gases. It was remarked that no differences were noted in pulmonary function due to smoking habits or type of welding processes and materials to which welders were exposed (for example, stainless steel welding or metals coated with anticorrosion primers).

Although there was a significantly greater prevalence of simple chronic bronchitis in welders than in controls, persistent cough and breathlessness were more frequent complaints among the controls than among the welders. There were no differences in rates of simple chronic bronchitis due to age, smoking habits, duration of weld-

Table 10
Chronic respiratory symptoms with and without lung dysfunction by smoking habit and occupation

	Welders								Unexposed controls							
	Smokers (N=71)				Nonsmokers (N=29)				Smokers (N=59)				Nonsmokers (N=41)			
	Normal lung function		Lung dys-function		Normal lung function		Lung dys-function		Normal lung function		Lung dys-function		Normal lung function		Lung dys-function	
Symptom	N	%	N	%	N	%	N	%	N	%	N	%	N	%	N	%
Cough	6	9	15	21	3	10	1	3	8	14	2	3	0	0	1	2
Expectoration	4	6	10	14	3	1	1	3	5	8	2	3	0	0	1	2
Dyspnea	4	6	3	7	2	7	1	3	1	2	1	2	0	0	0	0
Wheezing	2	3	10	14	1	3	0	0	1	2	1	2	0	0	0	0

Hunnicutt et al., 1964 (Ref. 9).

ing exposure, or welding processes and materials used.

The mean values of pulmonary function tests were similar for all welders with or without chronic bronchitis (Ref. 7).

Dreessen et al., 1947 (Ref. 3), found an excess of upper respiratory tract abnormalities (such as nasal congestion, sinusitis, and pharyngitis) in their study of over 3,000 welders and 1000 nonwelders employed in seven United States shipyards in steel ship construction. The excess of upper respiratory symptoms was especially noteworthy in male welders who smoked. In females, the difference between smokers and nonsmokers who welded was less pronounced than in the males. In a breakdown of data according to race and occupation,

Table 11
Chronic respiratory system clinical and functional variables in sheet metal welders

Variable	All welders	Controls	P	Welders with less than 20 years exposure	Welders with 20 years or more exposure	P
Number	25	20		12	13	N.S.
	Mean (range)	Mean (range)		Mean	Mean	
Age	48.8 (25-70)	46.7 (25-64)	N.S.	54.0	54.7	N.S.
Duration exposure (yrs.)	18.7 (3-32)	---	---	---	---	---
	N (%)	N (%)		N (%)	N (%)	
Smokers (1 pack per day for 5 yrs.)	14 (56%)	11 (55%)	N.S.	6 (50%)	8 (61%)	N.S.
Chronic cough	2 (8%)	2 (10%)	N.S.	---	---	---
Dyspnea	0 (0%)	0 (0%)	N.S.	---	---	---
Wheezing	3 (12.0%)	0 (0%)	N.S.	---	---	---
Chest x-ray, siderosis	8 (32.0%)	0 (0%)	<0.01	2 (16.7%)	6 (46.2%)	N.S.
	Mean ± s.e.	Mean ± s.e.		Mean ± s.e.	Mean ± s.e.	
FEV (% predicted)	87.0 ± 2.8	91.6 ± 3.4	N.S.	89.3 ± 2.7	84.8 ± 5.6	N.S.
$FEV_{1.0}$ (% VC)	75.1 ± 2.1	72.9 ± 2.7	N.S.	77.3 ± 3.3	73.2 ± 2.9	N.S.
RV (% predicted)	86.4 ± 3.6	90.1 ± 5.7	N.S.	---	---	---
TLC (% predicted)	83.1 ± 1.9	88.0 ± 2.4	N.S.	---	---	---
RV/TLC (% predicted)	100.8 ± 8.4	99.6 ± 5.8	N.S.	---	---	---

N.S. = Not statistically significant.

Kleinfeld et al., 1969 (Ref. 10).

Table 12
Prevalences of chronic respiratory clinical and functional abnormalities in welders in Finland

Variable	Welders, % (N=157)	Controls, % (N=108)	P
3-4 yrs. exposure to welding	10	---	---
5-14 yrs. exposure to welding	51	---	---
15 or more yrs. exposure	39	---	---
Persistent cough	9	10	N.S.
Dyspnea	29	33	N.S.
Simple chronic bronchitis	24	14	<0.01
Mucopurulent chronic bronchitis	5	5	N.S.
	Mean ± S.D.	Mean ± S.D.	
FVC (liters)	4.95 ± 0.8	4.83 ± 0.8	N.S.
$FEV_{1.0}$ (liters)	3.96 ± 0.7	3.94 ± 0.8	N.S.
100 x FEV/FVC	80.0 ± 6.8	81.4 ± 7.5	N.S.

N.S. = Not statistically significant

Antti-Poika et al., 1977 (Ref. 7).

Table 13 presents the percentages of workers with positive findings.

No permanent disability was reported in the welders, although the prevalence of a so-called upper respiratory symptom complex was considered by the authors to be statistically and clinically significant. The nasal congestion and pharyngitis were stated to have been caused by exposure to welding fumes and gases.

The prevalences of cough, chronic bronchitis, exertional dyspnea, wheezing, and respiratory system abnormalities in welders employed at the Galati shipyard in Romania were studied by Barhad et al., 1975 (Ref. 11). From a total of 700 welders, 153 males were chosen: all with at least 5 years of welding experience; no evidence of major chronic chest wall, heart, or pulmonary diseases; and no previous exposures to substances considered to be respiratory hazards. The arc welders worked in large shops and confined spaces. Total dust in the welders' breathing zones was 6 to 36 mg/m^3, but levels as high as 151 mg/m^3 were found in confined spaces; nitrogen oxides ranged from 1.1 to 1.7 mg/m^3, carbon monoxide from 6 to 17 mg/m^3; and manganese fumes ranged between 0.6 and 3.3 mg/m^3, depending upon process and materials. Shielded metal arc welding with basic covered electrodes, flux cored arc welding, and oxygas welding were utilized. A group of 100 controls was chosen from among maintenance workers and turners in areas free of welding fumes and gases. They

Table 13
Percent prevalences of chronic respiratory abnormalities in U.S. shipyard welders

	Welders				Nonwelders			
	Total	White males	Non-white males	White females	Total	White males	Non-white males	White females
Upper respiratory tract abnormality	40.5	47.7	37.9	30.3	30.9	36.4	32.6	22.4
Nasal congestion	18.6	22.2	18.2	13.5	14.2	17.2	15.9	9.3
Pharyngeal irritation	24.4	29.9	31.5	15.4	16.6	21.1	19.6	9.3
Upper respiratory tract symptoms	31.8	36.1	27.3	26.5	21.3	24.3	17.4	17.3

Dreessen et al., 1947 (Ref. 3).

Table 14
Prevalences of chronic respiratory signs and symptoms in Romanian shipyard welders

Criteria	Welders (N=173)	Controls (N=100)	P
Cough	38 (22%)	14 (14%)	N.S.
Chronic bronchitis	34 (17.7%)	13 (13%)	N.S.
Dyspnea (during level walking)	34 (19.7%)	5 (5%)	<0.001
Wheezing, asthma attacks, or both	27 (15.6%)	7 (7%)	<0.05
Positive physical findings	24 (13.9%)	12 (12%)	N.S.

N.S. = Not statistically significant

Barhad et al., 1975 (Ref. 11).

suffered no chronic pulmonary diseases and were matched by age and smoking habits with the welders.

For the welders, the mean age was 34 years. There were 81 smokers (47 percent), 59 nonsmokers (34 percent), and 33 exsmokers (19 percent). They were given a questionnaire to determine the frequency of various respiratory system complaints, which were reported as in Table 14. There was a significantly higher prevalence of dyspnea and wheezing among the welders than among the controls. When the groups were separated by smoking habits, the results revealed most striking differences between exsmoking welders and controls, as shown in Table 15. In this group, the occurrence of symptoms might have been an inducement to stop smoking.

Chronic bronchitis occurred 1.5 times more frequently in welders than in controls. When smoking habits and age were considered, the analysis shown in Table 16 was provided.

The difference in prevalence of chronic bronchitis in the over 40 age group in the welders and the controls was close to the 5 percent significance level (borderline significance).

An evaluation of lung function, on the basis of smoking, was accomplished for 125 welders. Results are given in Table 17.

Of the 125 measurements, 25 (20 percent) showed ventilatory impairment, defined as an $FEV_{1.0}$ below 81 percent of predicted, or an $FEV_{1.0}/VC$ ratio below the age-specific limits of normal, or both. The influence of smoking on lung function was regarded as minimal in the welders. Twenty-one of the 25 were of the restrictive type and 4 were considered obstructive impairments.

Respiratory disease symptoms and lung function in 156 welders were studied by Fogh et al., 1969 (Refs. 12 through 14). The welders, who ranged in age from 30 to 60 years, all had more than 5 years of welding experience in shipyards or as engine-, boiler- or tankmakers, and were exposed most often to fumes from the welding of mild steels. A control group consisted of 152 randomly selected employees of the same plants as the 156 welders, but who worked in areas where welding was not normally performed.

The prevalence of chronic bronchitis was similar in the welders and the controls. The data are presented in Table 18.

Smoking habits were identical for the welders and

Table 15
Prevalence of dyspnea during level walking in Romanian shipyard welders

	Welders	Controls
Smokers	24%	6%
Ex-smokers	33%	6%
Nonsmokers	7%	3%

Barhad et al., 1975 (Ref. 11).

Table 16
Percent prevalence of chronic bronchitis in Romanian shipyard welders

	Welders	Controls
Nonsmokers	12	3
Smokers	26	21
Over age 40	21.5	9.8

Barhad et al., 1975 (Ref. 11).

Table 17
Evaluation of lung function in Romanian shipyard welders

	Nonsmokers	Ex-smokers	Smokers
Number	40	24	61
% of predicted forced vital capacity (FVC) (mean)	87.2%	85.4%	87.7%
% of predicted one second forced expiratory volume ($FEV_{1.0}$) (mean)	91.1%	90.8%	88.9%
100 x $FEV_{1.0}$/FVC (mean)	80.1%	80.5%	78.2%

Barhad et al., 1975 (Ref. 11).

the controls. Mean values for one second forced expiratory volume ($FEV_{1.0}$) were similar for both groups: 3.7 ± 0.68 liters for welders and 3.7 ± 0.77 liters for controls. Both means were 102 percent of predicted $FEV_{1.0}$ levels. However, the welders showed increasing $FEV_{1.0}$ impairment with increasing use of tabacco. The difference between welders and controls with identical smoking habits was not statistically significant. There were significant differences in $FEV_{1.0}$ between nonsmoking welders and light smoking welders ($p < 0.05$) and between nonsmoking welders and heavy smoking welders ($p < 0.01$). These differences were not noted when nonsmoking controls were compared to light or heavy smoking controls.

Table 18
Prevalence of chronic bronchitis symptoms in welders

Symptoms	Welders	Controls
Absent	104 (67%)	111 (73%)
Present	52 (33%)	41 (27%)
Total	156 (100%)	152 (100%)

Fogh et al., 1969 (Ref. 12).

Thirteen welders (8 percent) had obstructive pulmonary impairment while the impairment was questionable in 8. For the controls, 17 (11 percent) had impaired lung function, of which 12 were of the obstructive type, 5 were of the restrictive type, and 6 were questionable. There was no statistically significant difference between the welders and the controls in this regard.

Cough and sputum were proportionally related to tobacco use. There were no significant differences between smoking welders and smoking controls in relation to these symptoms. However, there were significant differences between nonsmoking and smoking welders ($p < 0.001$) and between nonsmoking and smoking controls ($p < 0.05$) in relation to the presence of cough and sputum. No increases in cough and sputum were noted for nonsmokers and exsmokers in relation to aging. Among both the welders and the controls who did not smoke, forced expiratory volumes were significantly higher than in the 2 corresponding groups of smokers.

In welders, but not in controls, respiratory symptoms increased with age: 25 percent of the welders under age 50 had symptoms, while 55 percent of the welders over 50 years old had symptoms. This difference was highly significant ($p < 0.001$). However, the groups studied were small, making the results difficult to evaluate, according to the authors (Ref. 12).

Peters et al., 1973 (Ref. 11), studied respiratory disease in welder and control populations. Sixty-one welders were selected from a shipyard where repair work was done. Exposures to fumes from coated and uncoated metals, low and high alloy steels, aluminum alloys, and bronze, from the use of SMAW, GTAW, and GMAW techniques were encountered in an intermittent "job shop" manner, in the shop and in confined (poorly ventilated) spaces on shipboard. A group of 61 pipecoverers was matched to the welders by age and duration of employment in the shipyard. Sixty-three pipefitters were selected as an unexposed control for the welders and the pipecoverers.

No siderosis was detected on x-ray films, and no other x-ray abnormalities that could be related to welding exposure were noted.

In all 3 groups, smoking habits were similar, as were

ages and heights. (Height is an important consideration in evaluating lung function studies.) Ten percent of welders and 9 percent of pipefitters had chronic cough. Three each of the welders and the pipefitters experienced shortness of breath. The results of pulmonary function tests were similar in the welders and the pipefitters: tests included forced vital capacity, one second forced expiratory volume, peak expiratory flow rate and volume, and total lung capacity and residual volume, among others.

In 36 smoking welders, the peak expiratory flow rate was significantly higher ($p < 0.05$) than in 25 current nonsmoking welders. The nonsmokers were further divided into those who never smoked (N=10) and exsmokers (N=15). The 10 who never smoked were approximately 10 years younger than the exsmokers. No ventilatory defects were noted in those who never smoked. Forced vital capacity and one second forced expiratory volume were considerably below expected values in the exsmokers, being similar to the results for the smokers.

Peters et al. noted that welders, pipecoverers, and pipefitters all had similar exposures to asbestos in the shipyard, which complicates comparisons of these 3 groups in terms of respiratory disease criteria. Furthermore, both the pipefitters and the pipecoverers could be exposed to welding fumes. Exposure to ozone and nitrogen dioxide, known pulmonary irritants in welding fumes, would be expected to contribute to airway obstruction, and pulmonary function tests would be expected to reveal a pattern of increased total lung capacity and residual volume. On the other hand, exposure to asbestos and other fibrogenic dusts would be expected to cause a decrease in total lung capacity and residual volume, characteristic of restrictive pulmonary disease. An increase in residual volume was noted in welders; it was proportional to the duration of exposure. This could have been a result of ozone and nitrogen oxide exposure during welding, or a result of smoking, or aging, or both. Either result tends to increase residual lung volume. The total lung capacity of the welders was in the normal range. In the asbestos-exposed pipecoverers, a decrease in total lung capacity was noted, which was partly consistent with expectations. Residual volume did not decrease.

In order to better determine whether shipyard exposure to welding fumes and asbestos affected pulmonary function, the same authors compared their original 61 welders, 61 pipecoverers, and 63 pipefitters with a group of 94 pipefitters (controls) from another shipyard involved in new ship construction. These men had minimal or no exposure to welding fumes or asbestos. This time, statistically significant differences were noted between the original group and the new controls for forced vital capacity ($p < 0.005$), one second forced expiratory volume ($p < 0.005$), and peak expiratory flow rate ($p < 0.05$). These three pulmonary function test values were lowest in the welders.

A comparison of the same three pulmonary function tests between the pipefitters and welders on the one hand and policemen who were matched by age and height on the other hand revealed considerably lower values in the welders on all three tests. The values for the policemen were not measured by the authors, but taken from recent literature. This was used as further evidence of the deleterious effect of shipyard welding exposure on the respiratory system. It should be noted that the degree of disability of the welders was subclinical, except for the presence of cough in 10 percent and shortness of breath in 3 percent, as described earlier (Ref. 8).

Ross, 1978 (Ref. 15), reported results of medical examinations of a group of 926 male manual metal arc welders in heavy engineering and shipbuilding occupations (no information on welding process, materials, or fume levels presented) and a group of 755 controls in other trades, excluding boilermakers, carried out over a 6-year period.

Past medical history revealed, both for welders who smoked pipes and cigarettes and for nonsmoking welders, a significant excess ($p < 0.05$) of pneumonia, bronchitis, and respiratory illnesses in older welders. No significant difference existed between the nonsmoking controls and welders in the amount of respiratory illness on the whole, or for pneumonia and bronchitis, with one exception: welders aged 50 to 59 had significantly more respiratory illness, pneumonia, and bronchitis than did the controls of that age group.

Productive morning cough was the only symptom that was significantly more prevalent in older smoking welders than in older nonsmoking welders (50 to 59 years old). Other symptoms, such as sore throat, catarrh, and dry cough, were equally prevalent in all welders regardless of age.

About 45 percent of the welders had been exposed to asbestos. No signs or symptoms of asbestosis were found, however.

Mean breathing test results for smokers and nonsmokers in each 10-year age group (20 to 60) indicated no significant differences between welders and controls for forced vital capacity, forced expiratory volume, or peak expiratory flow rate. Smoking welders in the 40 to 49 year age group had significantly greater ventilation restriction than nonsmokers of the same age group. Airway obstruction was noted in 24.4 percent of smoking welders and 20.3 percent of nonsmoking welders, on the whole.

Ross concluded that long term welding exposure caused a high incidence of respiratory illness in 50 to 59 year old welders who smoked. However, because many of the current welding processes are of relatively recent introduction, Ross felt it is probably too early to evaluate their adverse health effects, if any.

Wilhelmsen et al., 1977 (Ref. 16), investigated welders in the Gotaverken shipyards in Sweden, where most welding is performed indoors in an assembly hall or in small ship compartments. HT steel, containing 0.2 percent carbon, 0.1 to 0.5 percent silicon, 0.5 to 1.5 percent manganese, 0.5 percent phosphorus, and 0.05 percent

Table 19
Prevalence of chronic respiratory symptoms in Swedish shipyard welders

	Welders, %		Clerks, %		50-yr.-old men, %	
	Non-smokers (N=75)	Smokers (N=44)	Non-smokers (N=53)	Smokers (N=37)	Non-smokers (N=157)	Smokers (N=182)
Morning cough for 3 months a year	19	19	0	15	6	36
Phlegm day or night for 3 months a year	16	30	6	17	1	12
Wheezing apart from colds	19	30	4	21	6	13
Walking uphill dyspnea	37	36	10	38	21	25

Wilhelmsen et al., 1977 (Ref. 16).

sulfur, 0.25 percent chromium, 0.30 percent nickel, 0.10 percent molybdenum, and 0.35 percent copper, with a 95 percent zinc and chromium primer containing polyvinylbutural, phenol-resins, and iron oxide red was used. Electrodes were the basic covered type, with high potassium and calcium content. The fume concentration was stated to be high, yet was simultaneously claimed never to exceed 1 mg/m^3.

The mean age of the group of welders was 43.6 ± 9 years; the mean duration of exposure was 17.3 ± 8 years. For a control group of clerks, the mean age was 37.4 ± 5 years. A second unexposed control group, consisting of 339 50 year old men, was randomly selected from the local population.

The prevalence of cough, phlegm, wheezing, and effort dyspnea was ascertained in these three groups, according to smoking habits, giving the data in Table 19. Smoking welders complained of more symptoms than nonsmoking nonwelders. The difference was less marked when smoking welders and controls were compared.

The 119 welders were asked if welding, especially when high levels of smoke are present, caused cough, phlegm, wheezing, or dyspnea. The responses were as follows:

	Yes	No	Nonresponsive
Does welding in presence of much smoke cause cough?	73 (60%)	40 (39%)	6 (5%)
Does welding in presence of much smoke cause phlegm?	63 (52%)	50 (43%)	6 (5%)
Does welding in presence of much smoke cause wheezing?	36 (30%)	73 (12%)	10 (8%)
Does welding in presence of much smoke cause dyspnea?	48 (10%)	61 (82%)	10 (8%)

Table 20
Prevalence of chronic respiratory symptoms in welders in Sweden

Symptom	Male welders (N=134) N	Male welders (N=134) %	Males in other occupations (N=7,321) N	Males in other occupations (N=7,321) %
Chest pain walking uphill	22	16.9	1084	14.8
Intermittent claudication	17	13.5	930	12.7
Dyspnea walking uphill	34	26.8	1596	21.8
Sputum in morning	37	28.7	1706	23.3

Wilhelmsen et al., 1977 (Ref. 16).

Fifty percent or more of the welders thought there was an association between all symptoms and welding, with the exception of dyspnea.

Based on a cross-sectional examination of 7455 males in all occupations in Goteborg, representing a random population sample, Wilhelmsen claimed that welders experienced more respiratory symptoms than men in other occupations, but did not provide the results of statistical tests to support this claim (Table 20). The 134 welders in this sample worked in various industries and shipyards, under varying conditions (not specified). Overreporting of symptoms is one bias that Wilhelmsen et al. ascribed to, influencing their data.

Pulmonary function tests were compared between the welders and clerks who were matched by age, height, and smoking status. There were no significant differences between the welders and the controls with respect to $FEV_{1.0}$ (one second forced expiratory volume), VC (vital capacity), or flow-volume curves with flow measurements at different points in the vital capacity. However, there were significant differences in closing volume, closing capacity, and total lung capacity, as measured by N_2-dilution, between nonsmoking welders and matched nonsmoking clerks. The importance of these changes was not discussed.

Lung function was not found to deteriorate significantly among 23 nonsmoking welders who were tested on Monday before work (after a 2-day rest from welding), again on Monday after work, Thursday before, and Thursday after work (Ref. 16).

In a report by Ulrich et al., 1974 (Ref. 17), a clinical study was carried out in 80 shipyard welders, 58 of whom worked in semiconfined spaces (fume levels up to 340 mg/m^3 in the breathing zone) and 22 in big halls (fume levels up to 192 mg/m^3 in the breathing zone). There were 80 age-matched controls selected from the shipyard who were not exposed to welding fumes. The prevalence of respiratory symptoms and positive physical findings in these populations was tabulated as shown in Table 21.

Among the nonsmoking welders and controls, cough, expectoration, dyspnea, and nasal catarrh were evaluated as shown in Table 22.

Subjective breathing difficulties and irritation of nose and throat were more frequent in welders than in controls, although those working in confined spaces showed only slightly greater frequencies than those welding in assembly halls. Nonsmoking welders had a greater frequency of cough, expectoration, dyspnea, and nasal catarrh than did nonsmoking controls. Smoking welders had a higher frequency of dyspnea than did smoking controls. Furthermore, the effects of welding and smoking on dyspnea were additive, but the effects on cough and expectoration were not.

Results of lung function tests measuring vital capacity (VC) and one second forced expiratory volume ($FEV_{1.0}$) were similar for all welders and controls. However, significant decreases in $FEV_{1.0}$ were noted both for welders in enclosed spaces in comparison to controls ($p = 0.02$), and for welders in confined areas in comparison to assembly hall welders ($p = 0.05$). Chronic bronchitis was considered to be the cause of the lowered $FEV_{1.0}$ in the affected welders. Chest x-ray pneumoconiosis, present in 50 percent of the welders, influenced

Table 21
Prevalence of chronic respiratory tract signs and symptoms in shipyard welders

Symptom	Control (N=80)	Total welders (N=80)	Welders working in enclosed spaces (N=58)
Cough	29 (36%)	31 (38%)	24 (41%)
Expectoration	29 (35%)	30 (38%)	24 (41%)
Dyspnea	12 (15%)	25 (31%)	18 (31%)
Nasal catarrh	10 (13%)	13 (16%)	9 (15%)
Physical finding			
Lung wheezes	27 (34%)	29 (36%)	22 (38%)
Conjunctivitis	25 (31%)	52 (65%)	41 (71%)
Rhinitis pharyngitis	22 (28%)	52 (65%)	41 (71%)
Hypertension	5 (6.2%)	0	0

Ulrich et al., 1974 (Ref. 17).

Table 22
Prevalence of chronic respiratory symptoms in nonsmoking welders by exposure

Symptom	Control (N=25)	Total welders (N=32)	Welders not in confined spaces (N=12)	Welders in confined spaces (N=20)
Cough	3 (12%)	14 (44%)	5 (42%)	9 (45%)
Expectoration	3 (12%)	13 (41%)	5 (42%)	8 (40%)
Dyspnea	1 (4%)	6 (19%)	2 (17%)	4 (20%)
Nasal catarrh	2 (8%)	2 (6%)	1 (8%)	1 (5%)

Ulrich et al., 1973 (Ref. 17).

neither lung function nor respiratory symptoms (Ref. 17).

Two other studies are included for completeness. Pulmonary diseases in Chilean welders and nonwelders in the iron industry were analyzed by Benavides C., 1976 (Ref. 139). He studied the clinical findings of 69 arc welders and oxygas cutters and 69 age-, height- and weight-matched controls with similar clinical histories. The welders mainly used basic and acidic covered electrodes to weld mild steel, bronze, and other copper alloys. Work in semiconfined areas was performed by 93 percent of the arc welders. There were 28 smokers (at least 1 cigarette per day for 1 year), and 41 nonsmokers and ex-smokers. An initial analysis of the welders and the controls gave the information in Table 23. No x-ray evidence of siderosis or pneumoconiosis was found in any of the welders.

Sixty-two welders were questioned about chronic bronchitis symptoms: that is, dry or productive cough daily for at least three months per year for the previous 2 years. Chronic bronchitis occurred significantly more frequently in welders ($p < 0.05$), but neither age nor years of exposure could be correlated with frequency of chronic bronchitis.

Stanescu et al., 1967 (Ref. 140), carried out pulmonary function tests comparing 16 welders with x-ray evidence of siderosis to a group of 13 healthy unexposed men. These welders (mean age 40 ± 10 years) were randomly chosen from 70 welders employed at a metallurgical plant and were further selected on the basis of having over 7 years of arc welding experience. No data on welding process, materials, or fume levels were provided. The control group (mean age 36 ± 8 years) consisted of men not occupationally exposed to dust. The two groups were matched in height, age, and smoking habits.

Table 23
Respiratory function and prevalence of chronic respiratory disease in Chilean welders

Variable	Welders	Controls	P
Forced 1 sec. expiratory volume ($FEV_{1.0}$)	3.3 liters (102.5% of predicted volume)	3.6 liters (107.6% of predicted volume)	
Forced vital capacity (FVC)	4.4 liters (109.4% of predicted volume)	4.7 liters (111.9% of predicted volume)	
100 x $FEV_{1.0}$/FVC	74.9	75.5	N.S.
Chronic bronchitis	22 (35%)	7.7%	<0.05

N.S. = Not statistically significant

Benavides C., 1976 (Ref. 139).

Seven welders had slight exertional dyspnea, and 3 experienced productive coughing, although these findings could not be related to smoking habits. Lung volumes were within normal limits in all but one case, as was the $FEV_{1.0}/VC$ ratio. However, there was a statistically significant drop ($0.01 > p > 0.001$) in linear static compliance and functional compliance in the welders as compared to the controls, as shown in Table 24. These tests evaluate lung elasticity. Five of 7 welders, with functional compliance less than the lower limit of normal, complained of dyspnea; 2 of the 7 were smokers. Iron deposits in the lungs, or possibly associated pulmonary fibrosis, or both offers an explanation for the decreased pulmonary compliance in this small group of arc welders with siderosis. A follow-up study was proposed to detect a possible increase in functional impairment or appearance of overt respiratory disease with time.

Serum Enzymes

Relationships exist between degree of tissue damage to the lungs and respiratory passages and the blood levels of certain enzymes released because of this tissue damage. The extent and the duration of exposure to welding environments were found to influence serum enzyme levels. Studies of welders have revealed elevations of IgA immunologobulin levels, pulmonary lactic dehydrogenase (LDH) isoenzyme levels, and serum trypticase (TPE) activity in significant proportions of welders (Refs. 141 through 143).

Carcinogenicity

No study specifically designed to study welders has been reported. However, several investigations that included welders are available. The risk of cancer for welders depends upon several factors, including the materials to which they are exposed, the adequacy of skin and respiratory protective equipment used, and individual susceptibility, among others. The possible confounding roles of asbestos in lung cancer and mesothelioma (Ref. 144), of cigarette smoking in lung cancer, and of other occupational carcinogens must be considered in evaluating these studies.

General Cancer Mortality

Ott, Holder, and Langer, 1976 (Ref. 31), studied the cause of death in a cohort consisting of 8,171 employees of Dow Chemical Company's Midland, Michigan, facility as of 1954, among which were maintenance welders. Individuals exposed to asbestos or arsenic were excluded. The number of deaths in the welders due to malignant neoplasms was higher than expected when compared to the rate for the white U.S. male population, but the increase was not statistically significant ($p > 0.05$). The malignancies did not favor any particular site. No particular causes were identified in the welders. No information on types of welding or degree of exposure was presented.

Lung Cancer

A NIOSH survey of death records for 1,376 male welders and flame cutters over age 20 in Washington state for the years 1950 through 1971 revealed significantly excessive numbers of deaths for: all cancers of the respiratory system; primary cancers of bronchus, trachea, and lung; and all bronchus and lung cancers ($p < 0.05$). Deaths in welders were compared to all male deaths in Washington during this period. In addition, there were excessive deaths from cancer of the pharynx, but this was not statistically significant (Ref. 30). Types of exposures, smoking habits, and other relevant conditions could not be considered.

Menck and Henderson, 1976 (Ref. 32), reviewed lung cancer cases and deaths for 3,938 males aged 20 to 64 in

Table 24
Pulmonary mechanics in industrial welders

	Linear static compliance in l/cm H_2O		Functional compliance in l/cm H_2O		Elastic breathing in g.cm/ml		Specific compliance in l/cm H_2O/l of FRC*		Maximum inspiratory pressure in cm H_2O	
	Mean	S.D.	Mean	S.D.	Mean	S.D.	Mean	S.D.	Mean	S.D.
Controls	0.204	0.044	0.157	0.024	2.47	0.70	0.045	0.007	24.2	5.9
Welders with siderosis	0.157	0.035	0.120	0.035	2.72	0.75	0.040	0.010	23.4	6.2
P	0.01-0.001		0.01-0.001		>0.80		0.10-0.20		>0.60	

*FRC = functional residual capacity.

Stanescu et al., 1967 (Ref. 140).

Table 25
Summary of important cross-sectional epidemiological studies of chronic respiratory system effects in welders

Ref.	Workplace[a]	Process/ electrode type[a]	Metal[a] welded	Exposure years	Percentage with abnormal chest x-ray films	Statistically signif. excess of chronic lung disease symptoms	Statistically signif. impairment in lung function indices
30	---	---	---	---	---	Yes	---
15	Shipyard	---	---	---	---	Yes, in 50-59 yr. old	No
7	Shop, with some confined work	SMAW/basic	Mild steel	>3	0	Yes/No	No
16	Shipyard, confined	SMAW/basic	Mild steel and primed metal	Mean 17 ± 8	---	Yes, in nonsmokers	Yes, in nonsmokers
6 10	Manufacturing	FCAW	Stainless and other steels	Mean 20	32	No	No
8	Shipyard repair Confined	GMAW, GTAW GMAW	Cr, Mn, Ni alloys Mild steel, aluminum, stainless steel	---	0	---	Yes
9	Shipyard			>10	34	Yes, in smokers	Yes, in smokers
12 14	Shipyards, or engine and boiler makers	SMAW	Mild steel, 30% with anticorrosion primer	>5	---	No	No
3	Ship construction				1.9	Yes	---
11	Shipyards, large shops and confined spaces	SMAW/basic FCAW, Oxygas		>5		Yes	
17	Shipyard	GMAW	---	---	---	Yes	No

a. Often, other exposures were implied but not reported.

Los Angeles County during 1968 through 1970 and 1972 and 1973. A comparison population of over 31,000 white males aged 20 to 64 in Los Angeles, a 1-in-50 sample taken from Census data, was used to compute Standard Mortality Ratios (SMR) for 417 occupation and 215 industry codes per the 1970 U.S. Census occupational classification system. Welding appeared among occupations with a statistically significantly increased ($p < 0.05$) representation among lung cancer deaths compared to welder's representation in the Census. It could not be determined whether this elevated lung cancer risk was the result of occupational carcinogens, including asbestos or polycyclic aromatic hydrocarbons, or was due to smoking habits or other factors, such as air pollution.

Breslow et al., 1954 (Ref. 33), undertook a case control study of the influences of occupation and cigarette smoking habits on lung cancer. A total of 518 lung cancer patients from 11 California hospitals for 1949 through 1952 were compared to a control group of 518 patients admitted to these hospitals for conditions other than chest diseases or cancer. Fourteen of the 518 lung cancer patients were male welders with at least 5 years occupational exposure; 2 of the 518 controls were welders; this difference was significant at the 5 percent level. Based on their representation in the cancer population, the welders would be expected to account for 9 cases, whereas 14 were observed. This difference was statistically significant at the 5 percent level, after correction for variation in smoking habits. Breslow et al. concluded that welding exposure was a risk factor in the development of lung cancer. But since the number of welders in the sample was quite small, further studies of welding as a risk factor in cancer were recommended.

Dunn et al., 1968 (Refs. 34 and 35), carried out a prospective lung cancer mortality study of 14 occupational groups in California, including male welders and burners with at least 5 years of occupational exposure, all of whom were 35 or more years of age. The sources of the study population were union organizations that furnished lists of members who were then contacted by questionnaires from 1954 through 1957; an 85 percent response rate was achieved. Each subject was followed up for an average of 7 years: until his death, his 70th birthday, or the end of the study in December 1962. Death rates among the welders were then compared to rates for the entire study population of 68,153 males in 14 occupations in California (including the welders). The statistical analysis showed that the lung cancer death rate for welders and burners was not significantly greater at the 5 percent level than expected, after corrections for age and smoking habits were applied. Table 26 summarizes the data.

Digestive Tract Cancer

In a case-control screening study, Houten et al., 1977 (Ref. 145), found a larger number of welders than expected among admissions to a New York cancer research hospital with cancer of the stomach, between 1956 and 1965. However, when compared to a group of patients at the hospital engaged in clerical occupations, the difference was not significant at the 5 percent level.

A NIOSH survey of death records of 1,376 male welders and flame cutters over age 20 in Washington state over the years 1950 through 1971 revealed a statistically significant ($p < 0.05$) deficit in the number of deaths from cancer of the digestive organs and peritoneum in welders compared to deaths in all occupations in Washington during these years (Ref. 30).

Skin Cancer

Studies suggest that chronic overexposure of the skin to welding arc radiation can cause "sunburn" and related premalignant and malignant skin disease. Powell et al., 1968 (Ref. 146), and Hinrichs, 1978 (Ref. 147), noted that the skin cancer causing ultraviolet wavelengths, 315 to 400 nm, are present in the welding arc radiation (Appendix A).

Based solely on a dearth of reports to the contrary, it does not appear that welders have an elevated risk of skin cancer when reasonable precautions are taken. Two cases of probable welding radiation-induced cancer have been found. Haneke and Gutschmidt, 1976(Ref. 148), reported a case of squamous cell carcinoma in a 54-year-old arc welder with 18 years of occupational exposure. Precancerous skin changes described as keratosis developed into malignancies of the right nostril, which then metastasized to lymphatics, nasal septum, and face. Ultraviolet radiation from the welding arc was thought to be the carcinogenic agent, as no face protection was worn.

Roquet-Doffiny et al., 1977 (Ref. 149), described a 58-year-old oxygas welder with chronic actinic dermatitis, erythema, atrophic plaques, telangiectasis, and pruritic keratotic lesions. Eight tumors were removed over a 4-year period from his face and neck: 5 basal cell carcinomas and 3 squamous cell carcinomas. The exposure to welding arc radiation over a period of 30 years were stated to have caused these lesions.

Table 26
Lung cancer deaths in welders and burners

	Total study population	Welders and burners
Man-years	482,658	81,389
Observed deaths	4,706	49
Expected deaths	-	46.5

Dunn et al., 1968 (Ref. 35).

A NIOSH survey of death records for 1,376 male welders and flame cutters over age 20 in Washington state during the years 1950 through 1971 did not reveal a significant ($p > 0.05$) excess number of deaths from cancer of the bladder, prostate, liver, biliary passages, and gallbladder, brain, or nervous system (Ref. 30).

Eye and Vision

This section deals with eye and vision disturbances in welders. Since welders and welders' helpers are now provided with eye protection against metal spatters and arc radiation, it is surprising that even within the last five years shipyard and heavy engineering welders have been reported to experience approximately two incidents of eye injury, accident, or arc eye per worker per year (Refs. 38 through 40).

Eye Irritation

Eye discomfort can result from overexposure to welding radiation, ozone, nitrogen oxides, breakdown products from the action of ultraviolet rays on chlorinated hydrocarbon vapors in the welding environment, and fumes from fluxes containing fluorides (Ref. 37).

Bates, 1962 (Ref. 150), noted that eye discomfort, fatigue, and pain, which can result from exposure to intense visible welding radiation, is probably a result of prolonged contraction of the ciliary muscles that control the shape of the lens. Pain from retinal nerve overstimulation was ruled out because the retina contains no pain receptors.

Challen et al., 1958 (Ref. 36), studied eye and nose irritation and lacrimation in operators of GTA welding machines who had 2.5 years of experience at this process. Eleven female and 3 male operators, chosen from a labor force of 36 welders, were interviewed; 11 complained of periodic eye irritation and upper respiratory tract symptoms, while 3 operators were symptom free. Ozone levels ranged from 0.8 to 1.7 ppm in workroom air. After reduction of the ozone levels to 0.2 ppm, no further complaints were reported. Trichloroethylene, present at levels up to 238 ppm, and phosgene (< 0.1 ppm) in workroom air might have contributed to the eye irritation. The authors failed to discuss eye protection in use at the time.

Epidemiology of Eye Accidents, Including 'Arc Eye'

Reports of eye accidents and injuries in welders are often not specific enough to separate by type of hazard. This makes the statistics of eye hazards in welders difficult to assess. Many reported cases of "arc flash" are in fact eye conditions resulting from small wounds or infections of the conjunctivae (Refs. 150 and 151).

In one Swedish shipyard, 4000 welders were involved in 11,000 eye accidents per year, according to Tengroth, 1976 (Ref. 38). In another Swedish shipyard employing 3,000 workers, more than 7,000 eye injuries were reported yearly, of which about 30 percent were caused by exposure to ultraviolet radiation (Ref. 39). No mention was made of the use of eye protective equipment.

Ross, 1973 (Ref. 40), reported that 459 non-lost time injuries occurred in 400 heavy engineering welders in 1971; 43 percent were eye injuries, with 1/3 classified as "arc eye." Flying particles were a frequent cause of both notifiable (lost-time) and non-lost time eye injuries in these welders. Use of eye protection was not discussed.

Benson, 1943 (Ref. 45), reported that approximately 5500 cases of actinic photophthalmia (arc eye) were treated at the Norfolk Navy Yard in Portsmouth, Virginia, over an unspecified period of time. Of 871 cases that were further analyzed, sick leave was required in 18.

The frequency of eye complaints in a shipyard in Portland, Oregon, employing 85,000 men and 25,000 women in 1941, including 3800 male and 1650 female welders was reviewed by Rieke in 1943 (Ref. 43). During a 6-month interval, 40 percent of the cases treated in the shipyard dispensary of the Oregon Shipbuilding Corporation were for eye complaints in electric arc welders. In the 21 months following the institution of improved training and engineering controls, the number of eye treatments was still over 30 percent of all treatments, totalling 500,000 cases.

Dreessen et al., 1947 (Ref. 3), found that in arc welders randomly selected from seven U.S. shipyards the prevalence of conjunctivitis was slightly higher than in controls not exposed to welding but matched by wage, training, and physical exertion, as well as by sex and race. For approximately 3000 welders and 1000 nonwelders, the prevalences of conjunctivitis in females were 11 percent and 8.5 percent, respectively; for males the figures were 18.8 percent and 17.1 percent, respectively. The differences were not statistically significant at the 5 percent level.

Corneal and Conjunctival Injuries

Keratoconjunctivitis, also known as arc eye welders' flash, and actinic ray photokeratitis, is an acute, self-limited irritation and inflammation of the superficial structures of the eyes, caused by cumulative overexposure to ultraviolet radiation given off by the welding arc. Infrared radiation can also burn the eyelids and cornea (Refs. 41 and 42).

Wearing appropriate protection, such as flash spectacles, goggles, welding shields, and helmets, prevents the development of these superficial eye lesions. Shiras, 1966 (Ref. 152), felt that the combination of spectacles with sideshields and a helmet were necessary to protect the welder from his own arc as well as that of any neighboring welders. The author recommended that welders' helpers, supervisors, and anyone else in the vicinity of welding operations should wear flash spectacles with sideshields.

When a welding operation calls for a filter shade higher than No. 6 (for example, oxygen cutting, heavy gas welding, and arc welding operations at 30 A or greater) helmets instead of goggles are necessary to protect the skin of the face as well as the eyes against the radiation emitted.

The unprotected cornea, conjunctiva, and iris absorb the ultraviolet wavelengths between 175 and 400 nm emanating from the welding arc, preventing transmission to deeper structures of the eye (Ref. 43). However, unlike the skin, which tans and thickens in response to ultraviolet exposure, repeated exposure of the eye to potentially hazardous levels of UV does not result in increasing the protective capabilities of the cornea (Ref. 153). Pitts and Tredici, 1971 (Ref. 154), found that peak absorbance of the corneal and conjunctival tissue occurs at 265 to 275 nm. According to Powell et al., 1968 (Ref. 146), keratitis is produced by ultraviolet radiation of a wavelength peak of 280 nm; conjunctivitis at 296.7 nm.

The cornea and aqueous humor absorb almost all infrared radiation of wavelengths longer than 1400 nm. Shorter wavelength infrared radiation is conducted to interior structures of the eye and raises the temperature of interior tissue as well as that of the cornea itself. Heating of the iris plays a major role in cataract development (Ref. 153). Thermal burns of the cornea from welding arc radiation have not been reported to date in the literature, however.

Rieke, 1943 (Ref. 43), pointed out two misconceptions about ultraviolet keratoconjunctivitis: the injury is more likely to occur at night; and unguarded exposure to a single bright flash as the welding arc is struck can cause this condition. Neither statement is true. The inflammation results only from prolonged cumulative overexposure to the welding arc.

The severity of ultraviolet keratoconjunctivitis is dependent upon the distance from the radiation source and the angle of incidence of the ultraviolet rays striking the eye. The critical exposure time is not well defined due to differences in the intensity, which depends upon welding process and materials used (see Appendix A, "Radiation from Arc Welding"). In 20 cases, Cascini et al., 1966 (Ref. 44), noted that exposures to welding radiation were 20 minutes to 2 hours in duration. There is also individual variation in susceptibility of the eye to ultraviolet radiation, as in sensitivity of the skin to sunlight, according to Bates, 1962 (Ref. 150). He noted that dark-eyed persons are least likely to suffer ultraviolet keratoconjunctivitis.

Symptoms of ultraviolet keratoconjunctivitis begin approximately 4 to 12 hours after exposure and may continue for about 18 to 48 hours. In about 4 hours following sufficiently prolonged exposure to ultraviolet radiation, the cytoplasm of the corneal epithelial cells begins to swell, leading to superficial epithelial cell devitalization and desquamation, eosinophilic cell infiltration, and minute edematous blebs on and in the corneal epithelium. The patient may report blurred vision, lacrimation but no eye discharge, burning pain, photophobia, headache, and a feeling of sand or grit in the eyes. The eyelids become red and edematous, and the skin of the face may appear sunburnt (Refs. 37, 43, 146, 155, and 156). In many cases, it is difficult to distinguish this condition from symptoms of trauma, infection, or the presence of a foreign body in the eye (Ref. 150).

No residual injury to the cornea or conjunctivae remains. According to Benson, 1943 (Ref. 45), no permanent eye injury was noted in the over 3000 cases he treated. English, 1973 (Ref. 42), concurred. No other eye structures become involved because the ultraviolet radiation is completely absorbed by the cornea, iris, lens, and conjunctiva before penetration can occur to the retina. Treatment involves responding to symptoms (Refs. 37, 43, 150, 155, 157, and 158).

Cases of ultraviolet keratoconjunctivitis in oxygas and arc welders have been reported (Refs. 156 and 159).

Cataract and Other Lenticular Effects

Lens injuries, accommodation, and visual acuity in welders are discussed.

Epidemiology

Reporting of cataract cases is inherently biased: while retired workers may try to claim cataract pension benefits, the reporting of cataracts among pre-retirement age groups may jeopardize their employment opportunities (Ref. 46). Another bias is that clinicians cannot distinguish between fully formed cataracts caused by aging and those caused by overexposure to infrared radiation (Ref. 90).

In a review of available data, Edbrooke and Edwards, 1967 (Ref. 46), failed to uncover evidence of an increased risk of cataract in welders. They examined data on welders in France (Ref. 133), statistics for industrial cataract cases, and very early studies by Vogt (1912), Crookes (1914), Cridland (1921), Healy (1921), and Roberts (1921). They also stated that the Netherlands and Swedish Welding Societies concurred with the opinion that cataracts do not occur more frequently in welders than in the general population, but no data were presented.

Fogh et al., 1969 (Ref. 12), studied a group of 154 electric arc welders selected from shipyards and various industries manufacturing engines, boilers, and tanks. The majority welded mild steel and had welded for 5 years or more on a full-time basis. Ophthalmological examination failed to reveal abnormalities that could be attributed to welding arc radiation exposure. The use of eye protective equipment in these welders was not discussed.

Pathogenesis

Infrared radiation from the welding arc has the potential of causing opacities in the eye lens. In glassblowers, smelters, and furnace workers, cataracts have occurred after 10 to 15 years of chronic exposure to infrared radiation (Ref. 41). According to Hinrichs, 1978 (Ref. 147), infrared radiation wavelengths of 700 to 1400 nm are cataractogenic, as are ultraviolet wavelengths of 314 to 400 nm. Pattee et al., 1973 (Ref. 90), noted that in the United States filter lenses must meet ANSI Standard 287.1-1968 for maximum allowable infrared transmittance, thereby providing ample protection.

In welding, infrared radiation from the arc, the molten metal, and the heat-affected zone metal are felt by the skin as warmth (Ref. 150). The effects on tissues are due to the temperature rise caused by the absorption of the radiation. The degree of temperature rise in a tissue is a function of wavelength, heat conduction variables, exposure time, and total amount of energy delivered to the exposed part. Because the radiation is lower in energy, it probably does not participate in photochemical reactions (Ref. 41). The skin provides its own warning mechanism for infrared exposure by having a pain threshold below its burn threshold. The eye lacks this warning mechanism (Ref. 37). Cataracts can therefore be produced in the lens by prolonged exposures at an intensity insufficient to cause skin burns (Ref. 41).

The iris absorbs much of the infrared radiation, which raises the temperature of the local vasculature of the eye, resulting in a warming of the aqueous humor and the lens (Ref. 156). Cataract formation has been directly correlated to the amount of energy absorbed by the iris and transmitted to the lens (Refs. 41 and 153). The lens cannot dissipate heat due to its lack of vascularization. This allows heat injury to occur (Ref. 150). The result is slow opacification of this normally clear structure over a period of 10 to 30 years.

Cases

Lenticular opacities similar to "smelters' cataract" were described in a 53-year-old welder with 26 years of exposure to arc and oxygas processes as well as heat for 6 years. Infrared radiation was considered as a possible influential factor, according to Szafran, 1965 (Ref. 160). The use of protective eye equipment could not be ascertained.

In 2 cases of cataracts in welders reported by Minton (1974), accidental striking of an arc while eye protection was not being worn was blamed for the eventual appearance of the cataract (Ref. 46). The ages of these welders, duration of exposure, or welding processes used were not specified.

These reports are inconclusive evidence of welding as a cause of cataracts, since clinicians cannot distinguish between fully formed cataracts caused by aging and those caused by infrared radiation overexposure (Ref. 90).

Accommodation of the Lens and Visual Acuity

Kodama, 1952 (Ref. 161), reported that in 49 welders, accommodation of the lens was reduced after a continuous arc welding task. The type of welding and extent of exposure to light, fumes, and gases were unspecified. This symptom also occurs in cases of vitamin B_1 deficiency. In the welders, vitamin B_1 excretion in the urine was reduced after work, and a similar reduction was noted in mice kept in the work area where the welding was being carried out. The author stated that excessive exposure to ultraviolet radiation might have therefore influenced vitamin B_1 levels. The use of adequate eye protective equipment was not discussed.

Ross, 1978 (Ref. 48), found no long term effects in the welders' vision in a cross-sectional study. Tests for distant and near vision were administered to 926 male manual metal arc welders with heavy engineering and shipbuilding experience (no other welding exposure information given) and a control group of 755 nonwelders. No statistically significant differences were noted between vision test results in welders and controls except that 20 to 29-year-old welders had significantly better vision ($p < 0.05$) than did age-matched controls.

Dreessen et al., 1947 (Ref. 3), noted, in a study of over 3000 male and female arc welders and 1000 nonwelders in 7 U.S. Maritime and Navy-commissioned shipyards, that females had generally poorer vision than males of the same age, but welders had the same visual acuity as nonwelders when groups of the same age were compared.

These studies clearly indicated that visual acuity deteriorates with aging. Key et al., 1977 (Ref. 41), also supports this in the observation that workers over age 40 generally require more light to perform a task than younger workers (no particular occupation was emphasized). In the same vein, Hinrichs (personal communication) stated that as welders advance in years, they choose progressively lighter welding lense shades during work. This practice, incidentally, exposes the eye to more light, increasing the risk of eye injury from welding radiation.

Retinal Injuries and Vision Impairment

Visual field blind spots, color blindness, and delayed adaptability to darkness have been reported in welders. These conditions may indicate impairment of the function of the retina. Damage to the retina can be detected during an ophthalmoscopic examination of the fundus (back of the eyeball).

Images coming through the lens are focused upon the retina, a dark pigmented structure lining the eyeball, and transmitted to the brain. The retina is responsible for color and black-and-white vision, dark adaptation, and some visual acuity, although the lens is responsible for focusing the image. Scotomas, interruptions in the field of vision caused by nonfunctioning areas on the retina, can be detected by clinicians. Large scotomas appear to a patient as blind spots. The retina contains no pain-registering elements (Ref. 150).

Pathogenesis of Retinal Injury

Sliney and Freasier, 1973 (Ref. 153), stated that visible and near infrared radiation from 400 to 1400 nm presented a chorioretinal burn hazard to the unprotected eye. These wavelengths are transmitted through the cornea, aqueous humor, lens, and vitreous humor and are absorbed in significant doses, principally by the retina. Most are absorbed by the retinal pigmented epithelial cells and choroid. Thermal damage to retina and choroid is proportional to the duration of exposure to infrared wavelengths. Circulation of blood in the retina normally should dissipate the heat, preventing its buildup. However, the lens, which focuses visible light upon the retina, also can coincidentally concentrate infrared radiation on the retina (Ref. 41). Lesions resulting from retinal absorption of visible and near infrared radiation are clinically similar to those described in cases of senile macular degeneration (Ref. 162), as well as to eclipse blindness and focal retinitis in sun gazers (Ref. 156).

Bates, 1962 (Ref. 15), stated that the glare resulting from fluxes used in the welding of light alloys and nonferrous metals, although rich in yellow (sodium) and red (lithium) wavelengths of visible light, was an annoyance but not a hazard to the retina.

Filipiakowa, 1970 (Ref. 163), reported results of retinal examination of the eyes of 180 welders and 110 welders' assistants. Hyperpigmentation of the macular area of the retina was noted in approximately 30 percent of cases; in the majority, infrared radiation injury was thought to be causative. Slight edema of the macular region of the retina was present in one of 31 patients with evidence of acute conjunctivitis.

In the two cases of retinal burn in welders that are presented below, it can be noted that: (1) eye protection was probably not in use at the time of the incident, and (2) neither worker suffered permanent vision loss.

Naidoff and Sliney, 1974 (Ref. 47), reported a case of retinal injury in an 18-year-old arc welding student who had stared at a gas tungsten arc during welding of aluminum for approximately 5 to 10 minutes without protective lenses while standing at a distance of approximately 200 cm from the arc. Facial erythema and keratoconjunctivitis developed a few hours after exposure. A dense central scotoma and peripheral field constriction developed, corresponding to a deep foveal lesion, which was surrounded by edematous retinal tissue. Nine days later, visual acuity was essentially unimproved. Pigmented granules surrounding the foveal lesion became more numerous by the twelfth day after exposure. Twenty days after the incident, the scotoma cleared and vision returned to normal. Ophthalmoscopic examination 16 months following the exposure revealed a residual depigmented lesion in the retina, approximately 250 x 500 μ in area, with hyperpigmented borders. Other reports have described permanent lesions varying in severity from retinal pigment irregularities to retinal cyst or hole formation, according to these authors.

A case of retinal injury in a 29-year-old black male arc welder was reported by Ruprecht, 1976 (Ref. 164). Infrared radiation overexposure most likely caused the retinal lesion and resultant diminution of vision. The use of eye protection could not be determined. Systemic corticosteriods brought about a gradual improvement in vision in both eyes over a 2-month period following the exposure.

Color Vision

Color blindness is a sex-linked genetic disorder in which specific groups of color-receptive cone cells of the retina are missing from birth. Approximately 2 percent of all men are red color blind; 6 percent of all men are green color blind; 8 percent of all men are red-green color blind; and 0.4 percent of all women have red-green color blindness (Ref. 165).

Ross, 1978 (Ref. 48), noted that 84 (9 percent) of 926 male manual metal arc welders with heavy engineering and shipbuilding experience (no other exposure details given) had color vision defects. Ross stated that this prevalence was about that expected in the normal male population. Cascini et al., 1966 (Ref. 44), noted 2 percent green and 4 percent red color blindness in a population of 46 welders whom they examined.

Filipiakowa, 1975 (Ref. 166), found what she considered a high prevalence of red color blindness in eye examination of 180 arc welders and 110 welders' assistants. However, the number with color blindness was too small to draw significant conclusions. The same deficiency can be attributed to a study by Sevcik et al., 1960 (Ref. 53).

Dark Adaptability

Dark adaptability is the speed with which an individual regains vision when lighting is suddenly reduced in intensity, a function of the rods in the retina. This is not to be confused with night vision.

Although El Gammal et al., 1973 (Ref. 156), found no experimental evidence for impaired dark adaptability of the eyes from welding exposures, other researchers disagreed.

A study of dark adaptability of the eye as influenced by welding arc exposure was reported by Filipiakowa, 1971 (Ref. 167). In 25 arc welders and 22 welders' assistants, dark adaptability was considered normal. In 23 welders suffering from acute conjunctivitis, there were 6 cases in which dark adaptability was markedly slowed. Unprotected exposure to the welding arc radiation was suspected to have altered the rods of the retina. The alterations disappeared in 7 days in the tested workers.

At the Goteborg Clinical Neurophysiology Laboratory, a method was developed to study the glare recovery or readaptation time of the eye (RAT). RAT depended upon age, oxygen content in breathing air, and physical fatigue. An association between RAT and welding was postulated to be due to inhalation of welding fumes (Ref. 168). The nature of this association was not specified.

Nervous System

In this section, acute and chronic nervous system disturbances in welders will be reviewed, with major emphasis on loss of consciousness, manganese intoxication, and lead poisoning.

General Mortality

An occupational and cause-of-death survey of 300,000 Washington state male death records from 1950 to 1971 did not reveal a significant ($p > 0.05$) excess number of deaths in welders and flame cutters due to cancer of the brain and nervous system or cerebrovascular disease when compared to all white male deaths in Washington during these years (Ref. 30).

Loss of Consciousness

Loss of consciousness can be produced by brain disease, alcohol or drug overdose, many acute or chronic diseases, and trauma. Insufficient oxygen, excessive carbon monoxide, or carbon dioxide in breathing air (especially in welders working in hot, confined spaces without adequate ventilation or a respirator) can bring on unconsciousness either slowly or suddenly.

According to Ross and Parkes (Refs. 169 and 170), carbon dioxide shielded arc welding processes may place a careless welder at risk. High concentrations of CO_2 cause suffocation by exclusion of oxygen. Since CO_2 is colorless, odorless, and heavier than air, the gas can accumulate in confined spaces (for example, the base of a vessel) without its presence being noticed.

Asphyxiation does not usually cause functional brain damage if the victim is promptly resuscitated. Carbon dioxide does not damage the brain tissue; it simply excludes oxygen from the atmosphere.

Manganese Intoxication

Chronic manganism is a progressive disease of the nervous system caused by prolonged overexposure to manganese compounds. It has been reported only rarely in welders, who are at risk when concentrations of manganese oxide fumes and dusts of manganese flux in the workroom air become excessive. Appendix A discusses sources of manganese exposure. Rieke, 1969 (Ref. 171), noted that lead is also toxic to the brain and peripheral nerves, and intoxication presents a varied clinical picture. The diagnosis of lead intoxication in suspicious cases by use of chelating agents easily distinguishes it from manganism, however.

Epidemiology

Although a number of cases (Refs. 174, 176, 183, and 306) have been reported, there are no recent studies reported in the literature that explore the magnitude of the problem of chronic manganese poisoning in welders. In future epidemiological studies of various welding populations, the prevalence of this disease should be investigated.

Pathogenesis

Chronic manganese intoxication is a disease characterized by neurological and psychological disturbances and pulmonary abnormalities associated with inhalation of manganese dust in the lungs (Ref. 172). Vorontsova, 1949 (Ref. 173), noted that chronic inhalation of 50 to 110 mg/m^3 welding fumes containing 8 to 10 percent manganese oxides from the electrodes was toxic. The OSHA standard is 5 mg/m^3 of manganese (ceiling value). In trace quantities, manganese is an essential element; Hine, 1975 (Ref. 174), stated the human requirement for manganese is 3 to 9 mg per day, with the liver, pancreas, and bone containing the highest concentrations in the body.

The disease may be reversible in mild cases, and the psychological disorders tend to subside after exposure to manganese is discontinued. The neurological disturbances may progress to partial or total disability (Ref. 175).

Early symptoms of chronic manganism include restlessness, irritability, and a tendency to cry or laugh without purpose. These symptoms may be followed by apathy, visual hallucinations, uncontrollable impulses, flight of ideas, mental confusion, or euphoria.

Mask-like facial expression, spastic grin, muscle rigidity, slow gait with sliding of the feet, increased and abnormal reflexes, monotonous blurred speech with poor articulation, tremors, irregular handwriting, impaired hearing, double vision, abnormal reactions to pain, touch, heat, and pressure, excessive salivation and perspiration, sexual impotence and diminution of libido have been described by various authors (Refs. 53, 54, and 175 through 177). Mental activity is reported to be slowed, judgment impaired, and memory weakened, but intelligence remains normal (Ref. 172).

In manganism, there is usually a low white blood cell count and increased levels of manganese in the blood, urine, hair, and fingernails. Some illnesses that may be confused with manganism include Parkinson's disease, multiple sclerosis, paralysis agitans, advanced syphilis, Wilson's disease (progressive lenticular degeneration associated with liver degeneration), and epidemic encephalitis (Ref. 172).

The observation that manganism resembles Parkinson's disease deserves emphasis. Although no data on the prevalence of Parkinsonism in welders are available, there is a concern that some cases of manganese poisoning could be mistakenly diagnosed as Parkinson's disease. Further investigations may be warranted.

Manganism, like Parkinsonism, responds favorably to treatment with the drug levodopa (L-dopa), indicating that the two diseases may share certain biochemical abnormalities: depletion of dopamine (a neurotransmitter) in the basal ganglia of the brain; and depletion of melanin pigment content of the nerve cells of the substantia nigra, also in the brain (Ref. 172).

Diagnosis

Early detection of a potential danger to developing chronic manganism has been stressed by many authors. Platonov, 1976 (Ref. 177), suggested personality tests such as the Minnesota Multiphasic Personality Inventory (MMPI) as a tool for early diagnosis. In a study of 150 electric arc welders in shipbuilding industries in Russia, personality or functional changes occurred before organic clinical signs of manganism developed. EEG studies, on the other hand, have failed to reveal brain wave changes attributable to manganese intoxication (Ref. 172).

According to Hine, 1975 (Ref. 174), diagnostic tests

required to confirm a tentative diagnosis of manganism are: increased blood, urine, and cerebrospinal fluid levels of manganese; increased manganese excretion in urine following administration of calcium disodium EDTA, a chelating agent (Ref. 178); and reversal or improvement of the neurological signs following L-dopa treatment.

Elevated levels of manganese in urine, feces, blood, and cerebrospinal fluid can indicate a risk of developing manganism. Hine found that elevations in urine manganese levels (normal levels are listed in Table 27) correlated well with the severity of neurological findings in manganese intoxicated welders.

Table 27
Normal manganese levels in humans

Medium	Level	Ref.
Blood	2-8 μg/100 ml	174
Cerebrospinal fluid	0.4-1.2 μg/100 ml	174
Urine	0.1-0.8 μg/100 ml	174
Feces	1.6 mg/100 g	182

Pesaresi et al., 1968 (Ref. 179), investigated the risk of chronic manganese intoxication in arc welders using covered electrodes, some of which contain manganese oxides in the coatings. Blood and urine manganese levels were determined in welders before and after work, and levels did not exceed normal values in either situation, indicating no evidence for a manganese poisoning risk.

Ponomareva, 1957 (Ref. 180), found that arc welders with manganese intoxication excreted elevated levels of manganese in their urine, although values varied from time-to-time in exposed individuals. Manganese intoxicated subjects excreted 20-100 μg/l. In about 50 percent of healthy arc welders tested, all having 2 to 10 years of welding exposure, elevated urinary manganese levels were also noted. The numbers of welders were not specified.

Barborik, 1973 (Ref. 181), found that 50 welders using acid, or basic covered electrodes, or both and working in large factory halls excreted "negligible" manganese in their urine, and a mean of 4.6 mg of fecal manganese per day. No cases of manganese intoxication were diagnosed in these welders.

Jindrichova, 1973 (Ref. 182), studied fecal manganese excretion in 14 submerged arc welders exposed to atmospheric manganese levels below the Czechoslovakian industrial limit of 2 mg/m^3. Average duration of employment had been 11.4 years. From 42 fecal samples, a mean excretion of 3.58 mg Mn per 100 g of feces was noted. In a group of 27 unexposed controls, Mn excretion was significantly lower, averaging 1.6 mg/100 g ($p = 0.01$). Clinical and laboratory investigations failed to reveal any evidence of manganese intoxication, however.

In welders exposed to atmospheric manganese concentrations up to 125 mg/m^3, fecal manganese levels were 2.5 times higher than normal values (Ref. 183). The numbers of welders investigated and the processes used were not specified. These welders did not exhibit definite signs of manganese intoxication.

Lead Intoxication

The prevalence of lead intoxication (plumbism) in welders is unknown at present, but Rieke, 1969 (Ref. 171), noted that the vague symptomatology has frequently prevented its diagnosis. The disease is not restricted to central and peripheral nervous system manifestations; gastrointestinal disturbances, anemia, hearing loss, fatigue, and vague aches and pains may confuse the diagnosis in more mild degrees of intoxication (Ref. 171). More details are presented in the sections on blood, the gastrointestinal tract, the ear and hearing, and the musculoskeletal system. Diagnosis is expedited by the use of chelating agents such as calcium disodium EDTA in cases of suspected lead poisoning in welders who are exposed to sources of lead fumes and show mental dullness, fatigue, or nonspecific early signs of the disease.

Blood

Blood cell examination in employed welders, usually performed at the time of a physical examination, has not revealed abnormalities related to exposure in most reports. Various blood cell alterations, occasionally brought on by overexposure to fumes containing zinc, fluoride, copper, and lead, are noted.

A common laboratory finding in cases of metal fume fever is leukocytosis. The increase in white blood cells is transitory, usually lasting a day or two following the onset of the fever (Ref. 59). It is probably related to the release of a histamine-like substance from respiratory epithelial cells that have been damaged by inhaled metal oxide particles (Refs. 59 and 299) (see the section on metal fume fever).

Ross and Hewitt, 1976 (Ref. 52), found no significant differences in hemoglobin levels or white blood cell counts between a group of over 350 heavy engineering welders and 100 or more nonwelders. Similarly, blood cell counts were within normal limits for 20 arc welders exposed to zinc oxide fumes while working on galvanized metals in confined shipyard tasks and having an average of 6.6 years of welding experience, according to Chmielewski et al., 1974 (Ref. 60). Normal hemograms were reported for 402 arc welders with 5 or more years of experience, according to Marchand et al., 1964 (Ref. 49).

Dreessen et al., 1947 (Ref. 3), sampled blood from arc welders in steel ship construction in 7 United States shipyards. Hemoglobin content was no different for welders and nonwelders; nonwhites had consistently lower levels than whites, regardless of occupation. Female galvanized metal welders were found to be more anemic than females welding on non-zinc-coated metals; the difference was not evident in males, however. The authors also reported that 7.5 percent of all female welders and 3.8 percent of female nonwelders had lower red blood cell counts than expected, although the difference was not statistically significant.

Schuler et al., 1962 (Ref. 58), found no anemia or serum hemoglobin levels below 12.5 g/100 ml blood in examinations of 23 arc welders from various industries in Santiago, Chile.

Inhalation of welding fumes was stated to have caused mild blood dyscrasias in 10 young workers employed in confined arc welding for several years, according to Meneghini and D'Orofrio, 1949 (Ref. 184). Normochromic anemia (red blood cell count of 3.2 million) occurred in one case. Leukopenia, lymphocytosis, and occasional eosinophilia were noted in peripheral blood. Slight inhibition of granulocytic bone marrow activity was revealed in myelograms.

Challen, 1968 (Ref. 64), stated that chronic overexposure to fluorides can bring on a syndrome called fluorosis, characterized by anemia, thyroid or digestive tract disorders, or hardening of the bones and ligaments.

Lead poisoning, which has been documented in shipyard lead burners and welders exposed to toxic levels of lead fumes, is sometimes detected in blood smears. Basophilic stippling of red blood cells was considered suggestive of lead intoxication but not specific, according to Rieke, 1969 (Ref. 171). Lead also interferes with hemoglobin synthesis, which may result in anemia (Ref. 185).

Bataille, 1946 (Ref. 186), reported a case of hemorrhagic manifestations in a 35-year-old male welder exposed to copper fume (levels not reported). Purpura and stomach hemorrhages, as well as fever, headache, lower back pain, and weakness occurred. Erythrocytosis, leukocytosis, and a reduction in platelet count were revealed. The author briefly noted another case of an oxygas welder who developed gingival hemorrhages, although no further discussion of this was presented.

Genitourinary System

Available literature on kidney diseases in welders, most of which have developed following cadmium fume overexposures during welding, is discussed. Also included in this section is a report concerning the effects of occupational welding exposure on male fertility.

Kidney Disease

A NIOSH survey of death records for 1376 male welders and flame cutters over age 20 in Washington state during the years 1950 to 1971 revealed no significant excess number of deaths due to chronic nephritis when compared to the death records for all males in Washington during these years (Ref. 30).

Clack, 1975 (Ref. 118), stated that chronic cadmium exposure has probably been the cause of renal cortical necrosis, due to the toxic buildup of cadmium in this organ. Axford, 1975 (Ref. 112), noted the possibility of fatal renal cortical necrosis in cases of cadmium fume intoxication. In nonfatal cases, according to Beton et al., 1966 (Ref. 113), it remains to be determined whether or not permanent kidney damage results. The effects of cadmium are further discussed in the section on the respiratory tract.

Fertility and Potency

Haneke, 1973 (Ref. 187), surveyed male welders to determine the effect of this occupation on fertility. Of 61 arc welders studied, 57 were married and 51 of these had fathered children. In 2 cases where the couple was childless, the husband was sterile. Following studies including an andrological examination, sperm counts, sperm activity, ejaculum volume and fructose content, evaluation of libido and potency, and consideration of past medical history of mumps, nicotine use, alcohol consumption, venereal disease, and use of medications affecting spermatogenesis, Haneke concluded that welding could not be proved to influence male fertility.

Cardiovascular System

This section presents data on electrocardiogram (EKG) and blood pressure studies in welders. One series of reports (Refs. 50 and 51) describes an abnormal EKG finding in industrial steel arc welders in Italy. However, other studies from the U.S. (Ref. 9) and Europe (Ref. 49) have not identified any EKG abnormalities in welders. Blood pressure studies (with one exception) have shown conclusively that welders have either normal or slightly lower diastolic and systolic blood pressure than age-matched controls (Refs. 3, 48, 52, 54, and 62).

Attention should be given to the possible effects of carbon monoxide exposure. Carbon monoxide is formed by the decomposition of carbon dioxide used in gas shielded welding processes and also in the oxyacetylene flame (see Appendix A, "Radiation from Arc Welding"). Carbon monoxide avidly combines with hemoglobin, reducing the oxygen-carrying capacity of the blood. It has been shown to significantly ($p < 0.001$) reduce the time to exhaustion from exercise in healthy individuals breathing air containing 100 ppm for 1 hour (Ref. 188). Exposure for 4 hours to atmospheres containing 100 ppm shortens the time to the onset of pain during exercise in patients with ischemic heart disease and worsens their EKG (Ref. 55). It has been shown to aggravate peripheral arterial disease and encourage development of atherosclerosis in experimental animals (Ref. 56). CO levels are possibly associated with an increased fatality rate from myocardial infarction (Ref. 189), and, based on evidence in dogs and monkeys, it may increase the risk of ventricular fibrillation in those with heart disease (Ref. 57).

One case of arterial fibrillation and acute pulmonary edema has been reported in a shipyard welder who had been working in a confined space; the conditions were not described in detail (Ref. 190). Krivoglaz (Ref. 191) reported cardiovascular system abnormalities in welders using a carbon dioxide shielded process on manganese steel.

As discussed in the section on the respiratory tract, symptoms of acute pulmonary edema due to welding fumes and gases can mimic myocardial infarction symptoms of chest pain and tightness (Ref. 192). In addition, many chronic lung conditions, such as interstitial fibrosis, bronchitis, and emphysema, place an added strain on the heart as it labors to pump blood through thickened or damaged lungs. Right sided (pulmonary) heart failure may develop secondarily to these lung diseases. The heart at first enlarges, then gradually weakens until it is incapable of pumping blood rapidly enough to the lungs to prevent blood from backing up into the heart, resulting in heart failure (cor pulmonale). Heart disease can be screened by electrocardiograms (EKG). Some EKG studies in welders follow.

Electrocardiograms

In a study of 402 currently working arc welders in France, Marchand et al., 1964 (Ref. 49), detected no EKG abnormalities, and similarly negative findings were reported by Dreessen et al., 1947 (Ref. 3), in a study of 35 electric arc welders in large shipbuilding plants in the United States.

In a study of 58 industrial steel arc welders and 75 nonwelding healthy subjects in other professions, Franco et al., 1973 (Ref. 50), and Taccola and Franco, 1975 (Ref. 51), found differences in electrocardiograms between the two groups, as shown in Table 28. The Macruz Index is the calculated ratio of the duration of the P-wave to the PR-segment of the EKG. This ratio was significantly higher in all the metal arc welders than in the controls, regardless of age or duration of welding experience. However, there was no statistically significant correlation with lung function measurements or pulmonary diffusion capacity. This EKG finding was interpreted as being due to toxic changes in myocardial metabolism or circulatory dynamics occurring at the atrial level and possibly associated with pulmonary circulation. The regulatory mechanism responsible for atrial depolarization and atrioventricular conduction was affected, as demonstrated by the increase in P-wave duration relative to the PR-segment.

Blood Pressure

While some studies have revealed no difference in blood pressure between welders and nonwelders, others have found a lower blood pressure in welders. The fact that heavy engineering welders tend to be more physically fit than those in sedentary jobs may account for the latter finding.

In a study of 450 heavy engineering welders, aged 20 to 59 years, Ross and Hewitt, 1976 (Ref. 52), reported that there was no significant difference in blood pressure when they compared the welders to a group of 450 control subjects. There was no mention of how the sample or control group was chosen or other pertinent data. Rozera, 1966 (Ref. 62), noted similar findings in a study of 620 welders in Italy.

Ross, 1978 (Ref. 48), studied 926 male manual arc welders with heavy engineering and shipbuilding experience (no further exposure data provided) and 755 unexposed controls. He found a significant ($p < 0.05$) reduction in the mean systolic and diastolic blood pressures of 50 to 59-year-old smoking welders as compared to unexposed, age-matched smoking controls.

Dreessen et al., 1947 (Ref. 3), studied the health of arc welders from 7 United States shipyards who were selected randomly from a list of all 140,000 current shipyard employees, including 21,500 welders. The 3,234 welders chosen were compared to a control group of nonwelders, matched by wage, training, and the physical demands of the job. It was revealed that white male welders of all ages had lower systolic blood pressures than white male nonwelders. For white females, the same trend was present, although not as pronounced. In welders with 6 to 12 months of experience, blood pressure in two age groups, those younger than 25 years and those between 40 and 44 years, was lower than in those welders of other age groups. The slight lowering of blood pressure in arc welders was statistically significant compared to the blood pressures of nonwelders in these shipyards. Figures 1 and 2 illustrate mean systolic blood pressures according to age for welders and nonwelders of both sexes.

Table 28
EKG findings in industrial welders in Italy

	Age 20-30 yrs.	31-40 yrs.	41-50 yrs.	Mean
Macruz index for welders	2.04 (N=13)	1.99 (N=32)	2.02 (N=13)	2.01
Macruz index for controls	1.40 (N=27)	1.43 (N=29)	1.36 (N=17)	1.42
P	<0.001	<0.001	<0.001	<0.001

Taccola and Franco, 1975 (Ref. 51).

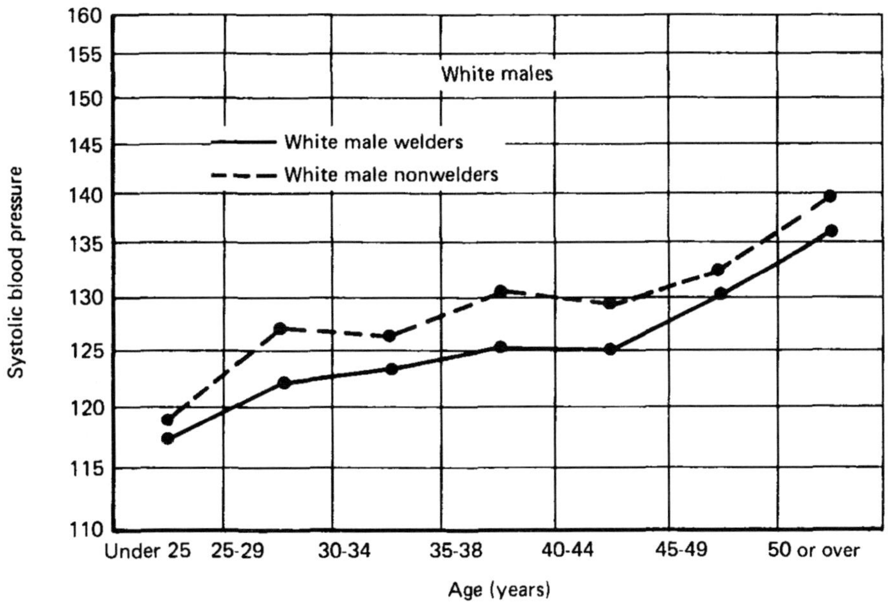

Fig. 1 – Mean systolic blood pressures of white male U.S. shipyard workers according to age and welding status

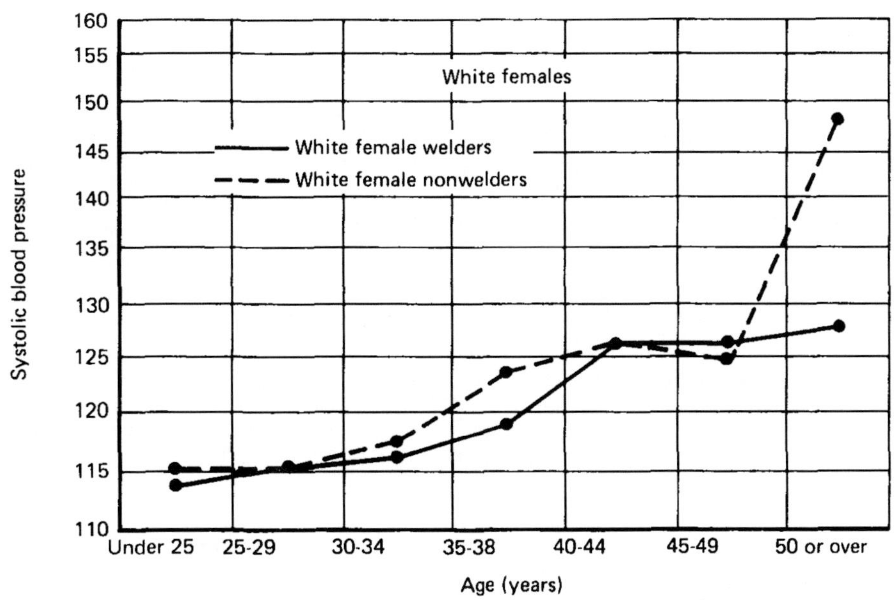

Ref: Dreessen et al., 1947 (Ref. 3).

Fig. 2 – Mean systolic blood pressures of white female U.S. shipyard workers according to age and welding status

Significantly lower mean blood pressure in arc welders employed at a northern Poland shipyard was reported by Walden-Galuszko and Kruminis-Lozowski, 1975 (Ref. 54). They chose 100 healthy workers 18 to 53 years old; there were 19 electric arc welders (welding techniques not specified) and 15 oxygas welders. The electric welders had a mean of 9.0 years of experience; the oxygas welders had 13.0 years of experience. The ventilation was considered poor for the electric welders but adequate for the oxygas welders, although no fume levels were provided. Blood pressure was significantly lower ($p < 0.05$) in the welders than expected age-dependent values. Pulse rates remained within normal limits.

Sevcik et al., 1960 (Ref. 53), noted that the systolic blood pressure of 53 arc welders employed in the iron and steel industry manufacturing metal containers was lower than expected. The welders were of mean age 36.6 years with an average of 8.6 years of welding experience. They worked in confined spaces, with exposures to manganese containing flux, copper, nickel and iron oxides, fluorosilicates, and silica (exposure levels not specified). The systolic blood pressures were lower than 110 mm Hg in 22 of the 53 welders.

Krivoglaz, 1973 (Ref. 191), noted a high frequency of arterial hypotension in welders using carbon dioxide shielded arc welding processes. No further information was provided.

Gastrointestinal Tract

The ingestion of welding fumes is one cause of gastrointestinal disturbances. In addition, it has been suggested that circulating metallic complexes originating in the lungs can induce toxic or allergic responses in the cells lining the gastrointestinal tract (Ref. 193). Future studies are necessary to confirm this theory.

Acute Effects

There is no firm evidence that welders have a greater risk of acute gastritis or other gastrointestinal symptoms than do nonwelders. Because metal fume fever and lead intoxication exhibit abdominal distress, some of the reported increases may reflect these conditions (Refs. 60 and 194). Hexavalent chromium fumes produced during the welding of stainless steel have been reported to be irritating to the gastrointestinal tract (Ref. 61). The following reports deal with the risk of acute symptoms in welders.

Gastrointestinal symptomatology in shipyard workers including welders was reported by Dreessen et al., 1947 (Ref. 3). The study involved examination of more than 3,000 welders and 1,000 nonwelders randomly chosen from then current employee lists of 7 United States shipyards. Five percent of welders and 4.5 percent of controls complained of symptoms referable to the gastrointestinal tract. Among white male welders, nonwhite male welders, and white female welders, 6.6 percent, 5.2 percent, and 2.9 percent, respectively, had gastrointestinal symptoms; for the control group, the figures were 5.7 percent, 0 percent, and 3.2 percent, respectively. Dreessen noted no difference in prevalence of gastrointestinal complaints between the shipyard welders and other shipyard employees.

A survey of acute symptoms in shipyard workers in Gdynia, Poland, revealed that welders, among others, more frequently reported acute gastrointestinal complaints (including diarrhea, nausea, and abdominal pain) after working hours than did galvanizers, according to Chmielewski et al., 1974 (Ref. 60). The study population included: 20 ship smiths having an average of 7.6 years of shipyard work; 20 arc welders working in confined spaces and superstructures, having an average of 6.6 years of experience; 20 engine room pipefitters working for an average of 10.7 years; and 20 galvanizers with much less exposure to zinc oxide fumes than the other three groups of workers. The results of the survey are shown in Table 29.

Table 29
Percent prevalence of gastrointestinal symptoms after work in a Polish shipyard

	Diarrhea	Nausea	Abdominal pain	Metallic taste in mouth
Arc welders	4	15	10	65
Pipefitters	11	21	20	75
Smiths	10	15	20	85
Galvanizers	1	1	1	80

Chmielewski et al., 1974 (Ref. 60).

These symptoms were the manifestations of metal fume fever due to high zinc oxide fume exposures at levels up to 1000 mg/m^3 in workroom air.

Three cases of subacute lead poisoning, as reported by Sklensky, 1969 (Ref. 194), occurred in oxygas welders working on the construction of a steel tower covered with lead-based paint. The exposure to lead vapors was said to be massive (no fume concentrations were given). The workers shortly developed abdominal cramps. The welders all recovered and returned to work following the episode.

Chronic Effects

Available reports from Italy and Russia disagree over the risk of ulcers and other chronic gastrointestinal diseases in welders. Many explanations are possible, including different exposures, work-related stress, diet, and predisposition to gastrointestinal ailments, among others. Studies on United States welders generally yielded no indication of an excessive risk from welding (Ref. 3).

Stancari and Amorati, 1963 (Ref. 193), reported on gastrointestinal diseases in a population of 264 arc welders from 23 industries in Bologna. No control population was studied. There was a high prevalence of gastric and duodenal ulcers and gastritis as years of exposure to welding (and age) increased: 67 percent of the welders working for 10 or more years had chronic gastrointestinal diseases. In many cases, a chronic erosive gastritis had developed. The etiology was stated to be uncertain.

Table 30 lists the prevalences of chronic gastritis, gastroduodenitis, and gastroduodenal ulcer in these arc welders.

Salamone et al., 1969 (Ref. 195), analyzed the prevalence of gastroduodenitis, and gastric or duodenal ulcers in 120 welders in metal machining industries in Italy who were hospitalized for various chronic respiratory and digestive system diseases. They were compared to 120 age-matched controls with no welding experience. No significant differences in findings on abdominal x-ray films were found, but selection of hospitalized patients may lead to bias.

Baranova et al., 1976 (Ref. 196), found that the prevalence of hypertension, cholecystitis, and gastroduodenal ulcer, as a group, increased with years of exposure to arc welding but depended only indirectly upon working conditions. A prevalence of 1.7 percent was given for shipyard arc welders.

Rozera, 1966 (Ref. 62), found increased morbidity from ulcerative and other digestive system diseases in a study of 620 welders in Italy's metallurgical and metal machining industries, as did Schuckmann, 1975 (Ref. 197).

Liver

There is almost no available literature in which liver disease or dysfunction in welders has been analyzed, the area evidently having been of minor concern to most investigators.

Certain serum studies can be accurate predictors of liver function. Some reports on liver protein synthesis and sugar metabolism, as revealed by serum levels of these substances, are presented in the following paragraphs.

Kny, 1942 (Ref. 198), reported that galactosuria, an indicator of disturbed hepatic sugar metabolism, was more prevalent in atomic arc welders than in aluminum or steel oxygas welders, as shown in Table 31.

The welders had occupational exposures of 7 to 22 years. No other exposure-related data were provided.

Kierst et al., 1964 (Ref. 132), studied 153 welders, employed at the Gdynia Shipyards in Poland, who were followed at the Outpatient Division for Occupational Diseases of the Institute of Marine Medicine. Ages ranged from 18 to 62 years; duration of welding experience was 1 to 32 years. As presented in Table 32, serum electrophoresis revealed changes in the numerical albumin-to-globulin ratio described as elevated albumin and decreased gamma globulins. The finding failed to correlate with either age or radiologic evidence of pneumoconiosis.

There was a directly proportional dependence of the two classes of serum proteins: the higher the albumins, the lower the gamma globulins. Albumin is known to carry Cu, Zn, Hg and possibly Mn compounds in the blood. These workers had exposure to iron, silica, manganese, cadmium, carbon and nitrogen oxides, and ozone, although levels were not provided.

Endocrine System

There is almost no information in the literature related to endocrinology in welders; what little is available does not indicate any particular concern for this aspect of the health of welders. One study of adrenal gland function in a small sample of welders is reviewed.

Mazza and Brancaccio, 1968 (Ref. 63), compared adrenal cortical and medullary function in a group of 25 welders and 10 unexposed controls of matched

Table 30
Prevalence of chronic gastrointestinal diseases in industrial welders in Italy

Welding exposure	Median age	N	Chronic gastritis, gastroduodenitis, or both	Gastro-duodenal ulcer	Total gastrointestinal morbidity
Up to 2 yrs.	22 yrs.	91	11 (12%)	2 (3%)	13 (15%)
3-10 yrs.	25 yrs.	99	41 (41%)	2 (2%)	43 (43%)
Over 10 yrs.	38 yrs.	74	28 (38%)	22 (29%)	50 (67%)

Stancari and Amorati, 1963 (Ref. 193).

median age. The welders were in the 21 to 46 year age range, with 5 to 15 years of welding experience at a large foundry complex in Italy where both arc welding and oxygas welding were used with various types of electrodes. No welders or controls had clinical evidence of endocrine dysfunction. As seen in Table 33, there were no significant differences in mean values of plasma cortisol, urinary corticosteroids, plasma or urine epinephrine or norepinephrine, or VMA between the welders and the controls. The Synacthen test is designed to measure the reserve capacity of the adrenal cortex through stimulation of cortisol secretion into the blood stream by administering the stimulant β 1-24 corticotropin. When welders and controls each received 0.25 mg of this stimulant intramuscularly, there were no significant differences in plasma cortisol levels between the welders and the healthy controls, indicating no evidence of adrenal cortex dysfunction in these welders. Table 34 shows the mean plasma cortisol levels obtained after the Synacthen test.

Table 31
Galactosuria in welders by process

Process	No. of welders tested	No. with galactosuria
Atomic arc	13	6 (48%)
Aluminum	12	4 (33%)
Steel oxygas	11	2 (22%)

Kny, 1942 (Ref. 198).

Musculoskeletal System

Chronic fluorosis and muscle fatigue are discussed.

Table 32
Albumins and globulins in blood serum of shipyard welders in Poland

	Albumins			Gamma globulins		
	55-65% (normal)	>65% (elevated)	<55% (decreased)	14-16% (normal)	>16% (elevated)	<14% (decreased)
Number	88	39	26	39	50	64
% of cases	57.5	25.5	17.0	25.5	37.2	41.8

Kierst et al., 1964 (Ref. 132).

Table 33
Adrenal function in foundry welders in Italy

	Welders		Controls	
	Mean	S.D.	Mean	S.D.
Plasma cortisol (γ%)	16.9	0.9	16.5	0.9
Total urinary corticosteroids (mg/24 hr)	6.5	1.67	7.3	1.39
Plasma epinephrine (γ%)	0.13	0.01	0.12	0.03
Plasma norepinephrine (γ%)	0.56	0.26	0.57	0.08
Urinary epinephrine (γ/24 hr)	5.8	0.36	6.1	1.58
Urinary norepinephrine (γ/24 hr)	35.0	10.1	39.5	9.1
VMA* (mg/24 hr)	5.18	0.51	5.16	0.55

*3 methyl-5-hydroxy-mandelic acid

Mazza and Brancaccio, 1968 (Ref. 63).

Table 34
Plasma cortisol (γ%) prior to and 30 minutes after synacthen test

	Baseline		After 0.25 mg i.m. β 1-24 corticotropin	
	Mean	S.D.	Mean	S.D.
10 controls	*16.5*	0.9	*30.8*	0.7
25 welders	16.8	0.8	31.2	1.0

Mazza and Brancaccio, 1968 (Ref. 63).

Chronic Flurosis

Fluoride ingestion has been associated with fluorosis of the bones and ligaments (Ref. 199). The lungs rapidly absorb inhaled fluorides, which are then cleared from the blood by the kidneys and excreted in the urine (Ref. 200). The bones, teeth, and hair retain fluorides to some extent. According to Challen, 1968 (Ref. 64), 1 ppm in drinking water is beneficial to dentition and in the prevention of dental caries. Chronic fluorosis is a syndrome characterized by increased density of bones and ligaments due to deposition of fluoride, possibly leading to pathological fractures, anemia, and digestive tract or thyroid disorders (Refs. 41 and 199). However, no literature is available that identifies any relationship between exposure to fluoride-containing fumes and disorders of bones or ligaments in welders (Ref. 64).

One possible reason why increased bone density has not been noted in the spine or ribs on welders' chest x-ray films was offered by Pantucek, 1975 (Ref. 200), who noted that a ferrous complex compound of fluoride is formed by the body. Its biologic activity is much lower than that of the fluoride ion alone. Aluminum and boron also form stable complexes that may lower the availability of fluoride. Welding fumes were stated to contain sufficient amounts of iron to fix all the fluoride present in the fumes, thus reducing the danger of adverse effects of fluorosis in persons occupationally exposed to fluoride-containing welding fumes.

Muscle Fatigue

Reports on muscle fatigue in welders presented in this section have been related to welding position (Ref. 201). Kadefors et al., 1976 (Ref. 202), Petersen et al., 1976 (Ref. 203), Peterson et al., 1977 (Ref. 168), and Ilner-Paine, 1977 (Ref. 204), reported that localized muscle fatigue was a common complaint in prolonged overhead work in Goteborg shipyard welders. Their study involved electromyography (EMG) of the back and shoulder muscles of these welders, of whom 10 were inexperienced (with less than one year on production welding work) and 10 were experienced (with more than 5 years of production welding work). The subjects had no history of chronic back or shoulder pain or appreciable trauma or disease.

Evaluations of low vertical welding, high vertical welding, and overhead welding were made. In each position, 4 weldings of about 2 minutes each were completed, the task requiring about 10 minutes to complete. In the inexperienced group, no discomfort was reported during low vertical welding, but fatigue occurred toward the end of the high vertical welding task. In overhead welding, many complaints of fatigue and pain were recorded. In the experienced welders, there were only a few complaints of fatigue towards the end of the overhead welding task. The EMG results showed that the inexperienced welders had significant ($p < 0.01$) muscle fatigue (modified EMG) in the supraspinatus and upper trapezius muscles during overhead welding, while for experienced welders, muscle fatigue was significant in the supraspinatus muscle alone. This indicated that complete adaptation to overhead welding positions does not occur and that overhead welding presents an undesirable working condition involving heavy static loading on the supraspinatus muscle.

Metal Fume Fever

This section discusses an acute allergic disease experienced by many welders during their occupational lifetimes. The frequency of incidents of metal fume fever in various welding populations and the small proportion of fevers that are complicated by pneumonia and pulmonary edema are noted.

Pathogenesis

Metal fume fever was recognized and described in 1832 as brass founders' ague. Manual arc welding was not performed commercially until 1910, however (Ref. 141).

Metal fume fever is known by many other names: galvo or zinc fever, braziers' disease, copper fever, Monday fever, the smothers, spelter shakes, and foundry fever (Refs. 183 and 185). Ross and Hewitt, 1976 (Ref. 52), and others (Refs. 118 and 183) noted that the most common cause of metal fume fever is overexposure to zinc fumes from welding, burning, or brazing galvanized steel. Cadmium fume poisoning mimics typical metal fume fever in its early stages, but cadmium is a much more severe, occasionally fatal, intoxicant (Ref. 118).

Metal fume fever is characterized by its acute onset (about 4 hours after exposure) and its usually short duration (about 24 hours). Recovery is complete; no permanent disability is suffered (Refs. 9, 118, and 183). The condition can mimic the flu, malaria, acute pneumonia, and upper gastrointestinal tract obstruction (Ref. 183). In general, chills, thirst, fever (102° F, 39° C), muscle aches, chest soreness, fatigue, gastrointestinal pain, headache, nausea, and vomiting may be the chief complaints.

The classical case of metal fume fever can be described

as follows: (It is emphasized that not all individuals experience all the symptoms, and severity is dependent upon the state of health of the welder prior to exposure, as well as to the degree of exposure and the degree of "temporary immunity" from other recent exposures.) Pharyngeal irritation, a metallic or sweet taste, and unquenchable thirst occur about 4 hours after exposure. Cigarette smoking is reportedly extremely unpleasant. By the end of the workday, nausea, lethargy, and dry cough may be present. A low grade fever, seldom higher than 102° F (39° C), develops after another 2 to 3 hours. One to four hours of chills and sweating precede the return of the temperature to normal. The patient then feels weak, and upper abdominal pain and repeated vomiting can occur (Ref. 183). Although the worker usually recovers in about 24 hours, a small percentage of cases may develop into acute pulmonary edema, or pneumonia, or both.

In a study by Volfovskaya and Makulova, 1971 (Ref. 205), 17 of 45 patients with metal fume fever developed acute pneumonia, which ran a rapid and typical course for an average of 15 days. In 24 of these patients, signs of pulmonary inflammation and edema occurred, lasting an average of 2 to 3 days. The patients were both arc and oxygas welders. The extent of exposure, materials, and processes were not specified.

Drinker et al., (1927), postulated that temporary immunity to further attacks of metal fume fever may be related to the presence of leukocytosis. Over the weekend, the white blood cell count drops to normal levels, permitting a recurrence of the attack after exposure on Monday (giving rise to the name "Monday fever") (Ref. 59). To date, there is still no consensus on the cause of metal fume fever or the observed temporary immunity it confers to welders (Refs. 59, 141, 183).

The treatment of the condition is supportive. Drinking milk may abate the nausea, vomiting, and gastrointestinal pain of metal fume fever, possibly by interfering with the reaction of zinc oxide with gastric hydrochloric acid, but milk ingestion does not prevent metal fume fever. The only prevention is avoidance of metal fume inhalation (Ref. 183).

Epidemiology

According to the results of studies by Chmielewski et al., 1974 (Ref. 60), zinc fever is relatively frequent in workers exposed to zinc oxide fumes. In a shipyard in Gdynia, Poland, 80 shipyard workers (including 20 ship smiths, 20 electric arc welders working in confined spaces and superstructures, 20 pipefitters working in the engine room, and 20 galvanizers exposed only to low zinc levels) were questioned as to whether they ever experienced fever, or chills, or both after work. The positive findings were as follows:

	Smiths	Welders	Pipe-fitters	Galva-nizers
Fever after work	35%	30%	35%	5%
Chills after work	35%	25%	50%	5%

All the smiths, welders, and pipefitters had experienced zinc fever in the past; it was noted that zinc oxide exposure was high among these workers and that they frequently ate and smoked while working.

Spacilova and Koval, 1975 (Ref. 129), compared the occurrence of metal fume fever in two groups of industrial arc welders in Prague who were matched by age and years of welding experience, but differed in that group A welders spent 13 percent of their yearly work time in enclosed and dusty environments while group B welders worked in open air for the most part. The results were as follows:

	Group A	Group B
Number of welders	23	14
Number and percentage reporting metal fume fever	22 (99%)	1 (7%)

Ross and Hewitt, 1976 (Ref. 52), questioned 530 heavy engineering arc welders about metal fume fever, finding that 31 percent stated they had experienced at least one attack. The most common event associated with occurrence of the fever was the welding of galvanized metal surfaces.

Ross and Parkes, 1975 (Ref. 169), personally questioned manual arc welders about exposure that could relate to metal fume fever occurrence. About 23 percent of 192 manual arc welders in heavy engineering and shipbuilding recalled having had metal fume fever symptoms and, upon questioning, attributed it to stainless steel welding, use of basic covered electrodes, welding galvanized surfaces, or welding on other materials.

Wilhelmsen et al., 1977 (Ref. 16), found that 69 percent of a sample of 119 male welders with an average of 17 years of experience responded positively when questioned about having had metal fume fever (with temperature 38 to 41° C). The exact figures were as follows:

Have you ever had metal fume fever?
- Yes: 83 welders (69%)
- No Answer: 2 welders (2%)
- No: 24 welders (29%)

These were workers at the Gotaverkin shipyard, Goteborg, Sweden, where a majority of welding was performed indoors in a large assembly hall or inside ship compartments. Basic electrodes were used on sheet metal primed with a paint containing iron, zinc, and chromium. Exposure levels were not specified.

In a study by Antti-Poika et al., 1977 (Ref. 7), a significant excess of fevers ($p < 0.001$) was found in a group of 127 male arc welders from shops in Finland, compared to an unexposed reference group of 93 males, corresponding to the welders with respect to age, smoking habits, and social class. There were 25 welders reporting a total of 54 incidents of fever, while 7 controls had experienced a total of 20 fevers. Some of both the controls and the welders had been exposed to dusts and irritating gases other than welding fumes for at least 1 year. The welders had been SMA welding with basic electrodes on mild unpainted steel in most cases, although some had also welded painted metals, or stainless steel, or both. The welders were exposed for at least 3 years for at least 3 hours per day. About 20 percent had worked in confined spaces for 25 percent or more of their employment.

Cases have been reported by Ross, 1974 (Ref. 141), Papp, 1968 (Ref. 183), Fishburn and Zenz, 1969 (Ref. 206), Paichl, 1964 (Ref. 207), Ilic and Popovic, 1971 (Ref. 208), Molfino, 1937 (Ref. 209), and Glass, 1970 (Ref. 210).

Ear and Hearing

Two problems of concern to welders are accidental metal burns of the ear and unprotected exposure to excessively noisy work environments.

Burns of the Ear

Van Petersen, 1971 (Ref. 211), reported a case of sudden deafness in a 22-year-old male welder. The condition resulted when a drop of hot metal penetrated the tympanic membrane. His hearing returned to normal in about one month after bed rest, vitamin supplements, and a vasodilator were prescribed.

Mobius, 1964 (Ref. 212), also described welding injury of the tympanic membrane. In examination of 193 arc welders, 25 percent showed scars and residual signs of perforation. The prognosis was considered excellent. Moller, 1975 (Ref. 213), reported similar findings in 7 Norwegian arc welders.

Welding spark injuries vary from minor burns to penetrating injuries leading to permanent inner ear impairment. Material imbedded in the tympanic membrane stimulates granulation tissue in the middle ear and, therefore, should be removed promptly, according to Andreasson and Elner, 1974 (Ref. 214).

Acoustic Trauma

Loss of hearing can result from continuous exposure to high noise levels. These levels are not precisely defined, since individual susceptibility plays a role in hearing loss. Other effects of noise exposure suggested by some authors (Refs. 118 and 215) include heart rate changes, fatigue, reduced attention span, loss of concentration, and lowered productivity on the job. Appendix A of this report discusses noise levels associated with various welding processes. OSHA's noise standard is 90 dB averaged over an 8-hour workday (also see Appendix B).

Ross, 1978 (Ref. 48), found hearing loss sufficient to affect speech communication (measured at 1, 2, and 3 kHz) in welders working in highly noisy areas, such as those near caulkers, where noise levels in excess of 90 dB are frequently present. The study included 926 male manual arc welders with heavy engineering and shipbuilding experience (use of ear protection not mentioned), and a control group of 755 nonwelders. Thirty-three welders had a "hearing loss of 34 dB or more" (increased threshold) in the better of the 2 ears in the speech frequencies (1, 2, and 3 kHz): one case was in the 30 to 39 year age group; 11 were 40 to 49; and 21 were 50 to 59 years old. An increase in hearing threshold in excess of an average 50 dB in the better ear in the speech frequencies occurred in one out of 220 welders in the 30 to 39 age group and in 3 out of 114 welders aged 50 to 59 years. The thresholds in the control population were not reported; the significance in welders was not discussed.

Acoustic trauma due to noise was reportedby Brusin et al., 1969 (Ref. 216). Oxygas welding noise levels were not considered excessive, but the mechanical-technical work areas in which the welding was performed registered noise levels above the permissible industrial limits for Yugoslavia. Health surveys revealed that 56 percent of the 75 welders tested by audiometry and otorhinolaryngological examination had occupational acoustic trauma; the frequency and degree of this finding were directly proportional to age and to years of exposure to the noisy environment.

Hickish and Challen, 1963 (Ref. 217), subjected three volunteers to noise from the plasma jet welding process for one hour. Audiometry results before and immediately after the noise exposure revealed increased mean hearing thresholds of +10, +20, and +35 decibels for frequencies of 1 to 2 kHz, 3 to 4 kHz, and 8 kHz, respectively. It was not possible to follow the hearing recovery pattern of all 3 subjects, but recovery of normal hearing required 48 hours in one volunteer. Sound levels up to 90 decibels in the octave bands of 0.6/1.2 kHz to 8/16 kHz were measured in the area normally occupied by the plasma torch operator.

Skin

The risk of developing skin conditions resulting from exposure to the sensitizing constituents of welding fumes, gases, ultraviolet and infrared light, and hot metal spatters must be reviewed in light of the protective clothing provided to the welder and the welder's decision to utilize this protection. The fully protected welder is at little risk.

Skin diseases are not unusually severe or frequent in welders, according to the results of many investigations (Refs. 3, 30, 52, and 155). Ross, 1977 (Ref. 155), noted that acne and heat rash were commonly diagnosed ailments in welders. Heat rash can be caused by the excessive warmth and moisture-retaining properties of some protective garments worn during welding.

Metal Burns

Skin burns resulting from metal spatter during welding leave small scars on the lower arm and other poorly protected areas of the body. Small metal burns may be slow to heal, according to Ross, 1977 (Ref. 155). Aluminum burns were said to be notorious for becoming infected.

Ultraviolet Radiation Overexposure and Photoallergy

The skin, as well as the eyes, can readily absorb ultraviolet radiation from the welding arc (Ref. 41). The severity of radiation injury depends upon such factors as protective clothing, welding process, exposure time, intensity of the radiation, distance from the radiation source, wavelength, sensitivity of the subject, and the presence in the body of skin-sensitizing agents that are activated by the radiation.

Reactions in humans exposed to ultraviolet radiation include erythema, skin cancer, urticaria, porphyria, photosensitivity, polymorphic photodermatitis, lupus erythematosus, actinic degeneration, photoallergic reactions to halogenated salicylanilides, drugs, and photosensitizing chemicals (Ref. 41).

Repeated exposure of lightly pigmented persons to ultraviolet radiation can result in an actinic skin condition: dry, brown, inelastic, and wrinkled skin. Actinic skin is not harmful itself, but it can develop into more severe conditions such as senile keratosis, squamous cell epithelioma, and basal cell epithelioma.

Photosensitizing chemicals (including furocoumarins and psoralens) as well as constituents of coal tar can cause exaggerated erythema and blistering when ultraviolet radiation exposure occurs. Some drugs, cosmetics, and foods (for example, figs, limes, and parsnips) carry these photosensitizers (Ref. 41).

Ultraviolet radiation-induced chronic dermatitis in a welder was reported by Balabanow et al., 1967 (Ref. 218). The 67-year-old male had been welding for 35 years, with dermatitis developing over the previous 2 years. The condition was clinically and histologically similar to lupus erythematosus, characterized by involvement of the face and extremities. The affected skin was hyperkeratotic with follicular hyperplasia, perivascular cell infiltrates, lymphocytes and histiocytes in affected areas. The use of skin protection could not be ascertained.

Schmitt and Silverman, 1971 (Ref. 219), reported that a 53-year-old welder, regularly employed at this trade for 3 years, developed chronic discoid lupus erythematosus (LE) when he switched from carbon arc welding to gas tungsten arc welding. The subject experienced burning on his chest, although he had worn a shirt, and a characteristic mottled rash on chin, forehead, ears, arms, chest, and face (presumably, these areas were unprotected during welding). Steroid administration gave immediate improvement. In the previous 10 years, he had infrequently used the GTA welding process and noted only slight transient skin erythema.

The association between chronic discoid LE and sunburn is well known (Ref. 219).

A case of occupational lupus erythematosus in a 27-year-old with no past or family history of photosensitivity reactions was reported by Gunche and Feoli, 1960 (Ref. 220). He had been arc welding with no head protection for 7 to 15 days at a time, including 8 hours a day on weekends. After 6 months he became aware of reddening and burning of the skin of his scalp, which developed over 3 months into typical discoid lupus lesions; alopecia and follicular hyperkeratosis were also noted. He stopped welding, and the lesions began to heal rapidly. It was felt that the ultraviolet and, possibly, the visible radiation from arc welding had caused the skin sensitization in an individual who was predisposed to the disease.

Wozniak, 1971 (Ref. 221), reported a case history of an arc welder with 13 years' experience who developed chronic discoid lupus erythematosus. There was no family history of skin diseases or allergies. The condition was described as eruptions, telangiectasia, follicular hyperkeratosis and a generalized increased skin sensitivity. Corticosteroids and avoidance of overexposure to ultraviolet radiation from the welding arc were effective treatment. The type(s) of welding performed by this worker and the use of protective equipment were not mentioned.

Skin Sensitization to Welding Fume Substances

Skin sensitizing substances, or irritating substances, or both in welding fumes, as well as other allergens present in the milieu (such as chromates, and compounds of nickel, zinc, cobalt, cadmium, molybdenum, and tungsten) produce dermatoses in shipbuilding welders, according to Szarmach and Synoradzka-Nakonieczna, 1970 (Ref. 222).

Hjorth and Wilkinson, 1969 (Ref. 223), noted that in Scandanavia and Finland, nickel was one of 20 of the most common contact allergens. Sensitivity to chromate was also important in Western European countries, occurring more often in men than women. Chromium and nickel compounds are found in fumes from stainless steel welding.

Gylseth et al., 1977 (Ref. 224), noted that chromium compounds can be corrosive to the skin and can cause hypersensitivity reactions. Hexavalent chromium compounds are usually responsible, although trivalent chromium compounds may also induce allergy. Tola et al., 1977 (Ref. 61), also noted the association between hexavalent chromium and allergic eczema.

Duengemann et al., 1972 (Ref. 225), tested 256 factory workers, the majority of whom were welders, for skin sensitivity to contact allergens that included calcium bichromate, cobalt, nickel, beryllium, cadmium, copper and zinc sulfates, and manganous chloride. Sensitivity to cadmium and copper was most frequently noted, possibly due to their presence in metal dusts and fumes. However, there was little sensitivity to Cr, Co, or Ni, and there were no noted reactions to the other substances tested.

Fumes from the welding of chromium steel can cause allergic dermatitis in persons sensitized to chromate. Epicutaneous testing showed that welding fumes contain allergenic chromium, according to Fregert and Ovrum,

1963 (Ref. 226). In one patient, welding with electrodes containing 0.1 to 1.0 percent chromium created fumes that induced facial dermatitis. Hexavalent chromium was said to be the cause. Job changes eliminated recurrent episodes of this hypersensitivity reaction.

Dermatitis was one symptom noted in zinc intoxication of shipyard welders working in confined spaces, according to a report by Molfino, 1937 (Ref. 209).

A case of urticaria and asthma in an oxyacetylene welder was reported by Kaplan and Zeligman, 1963 (Ref. 227). The 54-year-old worker suffered breathing difficulty and an itching skin eruption while engaged in preheat welding with a railroad rod using acetylene to weld battered rail ends out-of-doors. The rod employed contained mostly iron, with up to 0.4 percent carbon, 1.15 percent manganese, 0.04 percent phosphorus and sulfur, 0.55 percent silicon, 1.25 percent chromium, and 0.3 percent vanadium. The reactions had been occurring for about 2 years. When each exposure was terminated, the dyspnea subsided within one hour, and the itching was relieved in 6 hours. The reactions could be elicited by 10 minutes of exposure to the process. The authors felt that inhalation of gases and fumes emanating from the welding rod and acetylene flame were responsible, although no specific agent was identified.

Scleroderma

Fessel, 1977 (Ref. 228), reported a possible association between welding and scleroderma. Of 14 patients with scleroderma who attended the Kaiser Permanente Medical Center clinic in San Francisco, occupational histories revealed that 4 men and 1 woman had been welders at one time, 2 men being welders at that time. Three additional patients had indirect exposure to welding: 2 women had husbands who welded at home as a hobby, and one man worked 40 years as a machinist in close proximity to welding and had welded for one week, 30 years prior to the development of scleroderma. A control group of 54 male patients at the same clinic, with severe pulmonary disorders related to various occupational exposures, contained 5 former welders and 5 more with exposure to welding fumes and gases. There were 5 welders in the control groups, 5 of 54, as compared to 5 of 14 in the scleroderma group, a significant difference ($p < 0.05$). When the indirectly exposed persons were considered, accounting for 10 of 54 of the controls and 8 of 14 of the scleroderma group, the difference was also significant ($p < 0.02$). The role of welding fumes and gases in the pathogenesis of scleroderma warrants further investigation, according to Fessel.

Injury, Accident, and General Illness Statistics

A NIOSH survey of death records for 1376 white male welders and flame cutters over age 20 in Washington state during the years 1950 to 1971 revealed a statistically significant excess ($p < 0.05$) of deaths due to accidents caused by fire and explosion when compared to all white male deaths in Washington during these years (Ref. 30).

Ross (Ref. 40) studied a population of 400 heavy engineering manual arc welders over an 18 year period from 1954 to 1972 to determine the frequency and types of accidents that had been reported. There were 229 notifiable lost-time injuries, of which 18 percent were welding oriented. For 1971 alone, there were 6 non-notifiable lost-time accidents. The most frequent accidents for which time was lost were foot and eye injuries due to falling and flying objects, respectively. In 1971, there were an additional 459 non-lost-time accidents, of which 68 percent were related to welding. Eye injuries accounted for 43 percent, 1/3 of which were reported as arc eye; 19 percent involved fingers and thumbs; 10 percent involved hands and wrists. Flying particles caused 26 percent of the injuries; heat, 25 percent, and 15 percent involved injuries due to stepping on or striking against objects.

In 1964, Entwistle (Ref. 151) reported some statistics from the Engineering Inspector of Factories, Ministry of Labor, London, England. Oxygas welding accidents reported from January to September 1963 numbered 33. Eye injuries occurred in 31 cases, while 98 welding burns were noted. There were 5 welding accidents involving fires in 1961. A breakdown of reported electrical accidents for 1961 included one fatal and 27 nonfatal electric manual welding shock and burn cases, 3 nonfatal fixed welding apparatus electrical accidents, and 163 cases of electric arc eye flash. No estimate of the total workforce population was provided, and incidence rates were not tabulated.

Blood and Urine as Monitors of Welding Fume Exposure

Some relationships between blood and urine levels of various welding fume substances and occupational exposure to welding fumes have been investigated. Levels of iron, chromium, nickel, fluorides, zinc, manganese, and other elements in the blood of welders have been compared with blood levels in unexposed individuals. Excretion of fluorides, chromium, lead, nickel, and manganese in the urine of welders has been monitored. The results indicate a delicate balance in fume inhalation and lung deposition, absorption into the bloodstream, and excretion into the urine of healthy individuals.

Fate of Inhaled Welding Fumes

In conjunction with blood and urine studies, an understanding of absorption, distribution, and excretion of the substances in welding fumes is helpful. A simplified scheme of the fate of inhaled fume particles was illustrated by Ross and Hewitt, 1976 (Ref. 52). Four ways in which inhaled welding fumes are removed from the lungs and respiratory passages are: (1) by transport into the bloodstream; (2) by transport into the lymphatic channels; (3) via the gastrointestinal tract, by swallowing mucous containing the welding fume particles brought up from the lungs; and (4) by exhalation back into the air (which accounts for almost half of the inhaled welding fume particles). This scheme is presented in Fig. 3.

This figure represents a very simplified description of the physiological mechanisms involved; in fact, wide variation is expected due to solubility differences of fume constituents, particle size variation, and other factors. Figure 3 shows that only approximately 35 percent of fume particles are deposited in the pulmonary tissue, and that 40 percent of the deposited fumes are rapidly eliminated in the gastrointestinal tract. About 10 percent of the inhaled fume particles are deposited in the nasopharynx. In a half-time of 4 minutes, 90 percent of this portion may be eliminated via the gastrointestinal tract, and the remaining 10 percent enters the bloodstream. In the trachea, 8 percent of the welding fume particles are deposited; after a half-time of 10 minutes, 90 percent of this is eliminated via the gastrointestinal tract and 10 percent via the bloodstream. About 35 percent of the total welding fumes inhaled are deposited in the lungs, but after a half-time of 70 days, 15 percent of this amount is removed into the bloodstream, and another 5 percent by the lymphatics. Movement of welding fume particles from the lungs to the gastrointestinal tract accounts for 40 percent of the inhaled fumes within 1 day and another 40 percent in 70 days. The fate of welding fume constituents that enter the bloodstream is discussed in the following sections.

Blood Studies

Monitoring of blood levels of welding fume constituents absorbed mainly from the lungs into the bloodstream is reviewed.

Schuler et al., 1962 (Ref. 58), determined serum iron levels in the arc welders from various industries in Santiago, Chile. Levels were significantly lower in 23 welders with chest x-ray evidence of pulmonary siderosis than in 18 healthy control subjects: mean levels of serum iron were 97.1 ± 18.4 μg/100 ml blood and 119.7 ± 16.3 μg/100 ml blood, respectively.

Gillon and Marchand, 1963 (Ref. 133), reported elevated mean serum iron levels in 500 arc welders with at least 5 years of occupational exposure to fumes from shielded metal arc and gas metal arc welding. Serum iron determinations were unrelated to age or duration of exposure to welding.

Ross and Hewitt, 1976 (Ref. 52), analyzed concentrations of certain metal ions in the blood serum of two welders and two age-matched unexposed controls on 3 occasions at monthly intervals. No alterations in levels of iron, zinc, bromine, selenium, cobalt, or sodium were detected.

Ulfvarson and Wold, 1977 (Ref. 229), determined the concentration of 17 trace elements in the whole blood of 81 individuals working with different welding methods on stainless steel or aluminum and in 68 nonwelders as a reference group.

Zinc, manganese, chromium, silicon, and magnesium were assumed to be important constituents of the welding fumes for (1) SMAW of stainless steel, (2) GTAW of stainless steel, (3) GMAW of aluminum and aluminum alloys containing 1 to 5 percent magnesium, or silicon, or zinc, or combinations thereof, and (4) GTAW of aluminum and aluminum alloys. Aluminum and nickel could not be determined in the blood in this study. Iron in the blood probably did not reflect the exposure to iron oxide fumes because it is an important natural constituent of blood.

The welders studied worked 40 to 100 percent of

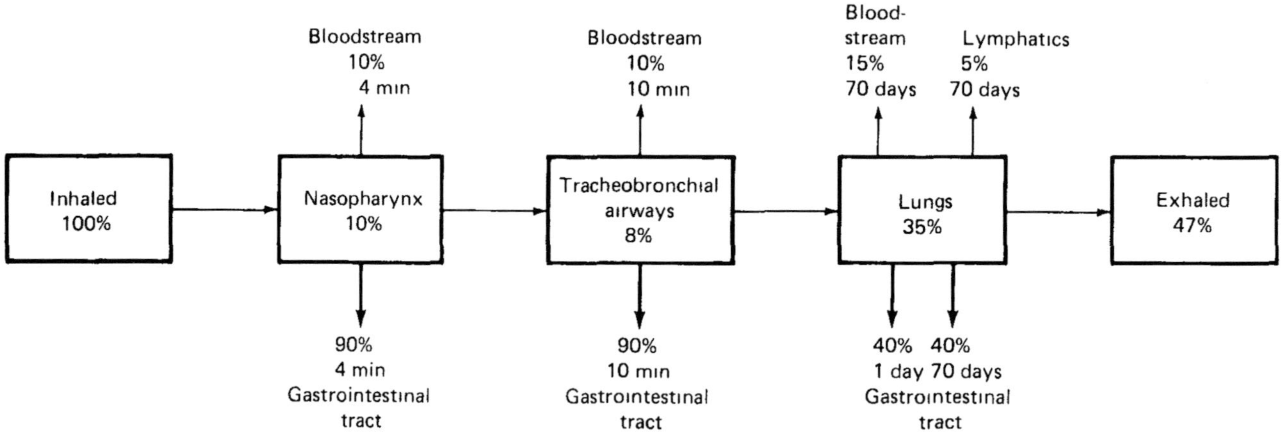

(Ross and Hewitt, 1976 (Ref 102).

Fig. 3 – Fate of inhaled welding fumes in humans

their shifts in the four aforementioned operations, with exposures to either 5 mg/m^3 (0.8 to 21 mg/m^3) stainless steel welding fumes containing 0.3 mg/m^3 CrO_3 and 0.03 mg/m^3 nickel fumes, or 12 mg/m^3 (0.3 to 150 mg/m^3) aluminum fumes. The concentrations were measured in the welders' breathing zone inside the mask on the day the blood samples were taken. Blood was taken at the end of the work shift.

Data were analyzed by the SIMCA computerized method for pattern recognition (that is, discriminant analysis). No differences were found between blood samples from welders exposed to the 4 different welding operations or control subjects, either in average levels of the trace elements or in correlational relationships between the trace elements. Thus, elemental analysis of blood samples did not reveal any effect of exposure of these subjects to welding fumes of aluminum or stainless steel by SMAW, GTAW, or GMAW processes (Ref. 229).

De Kretser et al., 1964 (Ref. 230), found that high levels of carbon monoxide (up to 300 ppm outside face shield; up to 70 ppm within face shield) were associated with elevated blood carboxyhemoglobin levels in 23 welders using a bare wire dip transfer CO_2 shielded process in a fixed jig area. Oddly enough, in a control group of 43 persons, as well as in the welders, smoking tobacco did not influence carboxyhemoglobin levels. A significant decrease in blood carboxyhemoglobin occurred after improvements in local ventilation were instituted. Although the atmospheric levels of CO after the improvements were not given, prior to the improvement, carboxyhemoglobin levels averaged 7.3 percent (range 0.5 to 20.2 percent), while after the improvement, carboxyhemoglobin levels averaged 1.4 percent (range 0.3 to 4.5 percent).

Urine Studies

Healthy kidneys are essential to the rapid clearance of many welding fume substances that enter the bloodstream. Conversely, kidney disease compromises urinary excretion; a toxic buildup of these substances in the bloodstream might then occur. As noted in the section on the kidney, cadmium fume overexposure can damage the kidney as well.

Fluoride

In a study by Ross and Hewitt, 1976 (Ref. 52), the excretion of welding fume compounds was assessed by analysis of urine samples from two welders, five hours after exposure to welding with basic covered electrodes. Urinary fluorides were 3 to 5 times greater in these welders than in age-matched unexposed controls. Challen, 1968 (Ref. 64), and Pantucek, 1975 (Ref. 200), noted that fluoride intake via absorption from the lungs usually is balanced by rapid excretion in urine, feces, and sweat, although some retention occurs in bone, teeth, and hair. Fluoride excretion in the urine can be monitored as an indicator of exposure because of its rapid excretion. Challen noted a lack of studies that combine atmospheric fluoride estimates with urinary fluoride levels in exposed individuals. Such studies would provide levels of excretion as a reference to degree of exposure, and might indicate occasions on which the TLV of 2.5 mg/m^3 for fluoride (Ref. 231) had been exceeded.

Pantucek noted that short term fluoride accumulation in the lung and slow release into the bloodstream was also possible, and that this was responsible for high levels of urinary fluorides occurring after termination of exposure. The urinary levels of fluoride in this case gradually decreased over an unspecified period of time (Ref. 200).

Krechniak, 1969 (Ref. 232), measured levels of urinary fluorides in 122 manual arc welders, 29 welding machine operators, and 10 controls. The welders were exposed to fluoride-containing fumes from either basic covered electrodes or automatic submerged arc welding processes. In a welding shop, average fluoride particulate levels and HF gas levels were measured after 150 min. of welding; the concentrations were 0.36 mg/m^3 of F in particulates and 2.3 mg/m^3 of volatile HF. After 2 to 5 hours of welding in a shipyard production hall, average fluoride particulate levels were 1.4 mg/m^3, while HF gas averaged 0.32 mg/m^3. In controls, the only sources of fluorides were drinking water and air pollution in Poland, where the study was done.

For the welders, urine samples were collected the morning following a regular work shift. In 46 percent of the machine operators, urinary fluoride levels exceeded normal limits, while this was true for only 6 percent of manual welders. As noted in Table 35, higher levels of urinary fluorides occurred in the machine operators. This was due to their higher exposures to atmospheric fluorides.

Chromium

Gylseth et al., 1977 (Ref. 225), evaluated chromium fume exposure in welders by determining levels of urinary chromium. Chromium levels normally vary from 0 to 50 μg/l of urine, depending upon geographical locations, nutrition, alcohol consumption, and analytic methodology as well. In workers exposed to hexavalent chromium fumes, urinary levels as high as 140 μg/l have been reported.

In this investigation, urine specimens from five welders exposed to steel containing 18 to 26 percent chromium and 8 to 20 percent nickel were collected before and after work on 5 subsequent days and compared to urine samples from unexposed subjects. Urine samples were analyzed for chromium by atomic spectrophotometry with a sensitivity of 2 μg/l. Before the work shift, mean chromium levels in urine were 8.8 μg/l (4.8 to 12.8 μg/l, 95 percent confidence limits). The second analysis, after work, revealed a mean value of 11.3 μg/l (7.2 to 15.4 μg/l). This second set of values was significantly greater than the first. Urinary chromium levels after work were directly proportional to workroom air total chromium levels ($r = 0.95$), and the correlation was highly significant ($p < 0.001$). The absorption and excretion of chromium were rapid processes, although the results did

Table 35
24-Hour urinary fluoride levels in welders

	No.	Duration of occup. exposure	Mean and S.D. of 24-hr urinary fluorides in mg/l
Manual arc welders	40	<5 yrs.	1.3 ± 0.88
	28	5-10 yrs.	1.4 ± 0.85
	54	>10 yrs.	1.2 ± 0.77
Automatic welding machine operators	29	Aver. 5 yrs.	3.2 ± 2.81
Controls	10	None	0.30 ± 0.16*

*Upper limit of normal = 2 mg/l in Poland

Krechniak, 1969 (Ref. 232).

not permit an estimation of the exact half-time of chromium excretion. Gylseth et al. assumed that most chromium compounds in welding fumes consist of soluble hexavalent chromium; since urinary chromium concentration of 40-50 μg/l immediately after work would correspond to 0.05 mg/m^3 of chromium fumes in workroom air, the present threshold limit value. However, see Appendix A for a detailed discussion of the chemistry of chromium welding fumes.

Tola et al., 1977 (Ref. 61), found that urinary chromium levels in 5 welders working with high alloy nickel-chromium steel (18 percent Cr) accurately indicated short-term exposure to soluble hexavalent chromium. When exposure was above 0.05 mg/m^3 of chromium, concentrations greater than 30 μg per gram of creatine were found in the urine. The 5 welders were healthy males, mean age 27.5 years, with 8.8 years of occupational exposure. For this experiment, which lasted one week, four of the subjects welded with rutile covered electrodes and one used GMA welding. A control subject welded mild steel with basic covered electrodes. The shop manufactured paper industry machines; ventilation was considered adequate although no local welding exhaust systems were in use.

The highest urinary chromium levels were usually found in urine samples taken at the end of each workday. Morning values were low, suggesting that chromium excretion had been rapid during the night. Over the weekend, however, the morning urinary chromium levels were slightly higher than those on Monday morning, suggesting that short-term accumulation of chromium was possible.

The concentration of water-insoluble trivalent chromium in workroom air failed to show any correlation to urine chromium levels. High levels of chromium were measured only during welding of high alloy steel with rutile covered electrodes, but not during GMA welding. The increase in chromium excreted during the workday reflected exposure levels better than measurements of afternoon urine chromium only.

If the chromium standard is lowered below 0.05 mg/m^3 for carcinogenic chromium according to NIOSH recommendations, urinary determinations would not be sensitive enough to indicate exposure, according to Tola et al. They recommended a study of the relationship between urine chromium concentrations and biological effects (Ref. 61).

Nickel

Jahr and Johnsen, 1974 (Ref. 233), studied nickel excretion in 18 welders exposed to stainless steel fumes. The level of nickel fumes in the breathing zone was directly related to levels of nickel excreted in the urine. The highest exposures occurred with electrodes containing approximately 80 percent nickel. The 13 welders using these electrodes were exposed to a mean fume level of 1.3 mg/m^3 (0.69 to 2.49 mg/m^3) measured over one week. In addition to nickel, the mean weekly chromium metal fume concentration in the breathing zone was 0.28 mg/m^3 (0.15 to 0.50 mg/m^3). The oxidation state of chromium in the fume was not determined.

Lead

Rieke, 1969 (Ref. 171), noted that monitoring lead excretion in morning urine samples was a simple and informative method of screening for lead overexposure in workers at risk. He surveyed the lead intoxication problem in welders and lead burners in shipbuilding and shipscrapping industries during and since World War II. No specific urinary lead values or exposure levels were noted.

Manganese

Manganese levels in urine are presented in the discussion of manganese intoxication in the section entitled "Nervous System." Manganese intoxicated subjects usually show elevated urine manganese levels. Hine, 1975 (Ref. 174), noted that urinary manganese levels correlated well with the severity of neurological abnormalities in manganism. Monitoring of urinary manganese may be useful in determining overexposure to these fumes in welding (Refs. 180, 181, and 183).

Chapter 3

Experimental Animal Studies

Exposures to welding fumes from many different types of electrodes and welding processes, mostly those utilized in Poland and Russia, are presented. Many studies are over 30 years old. The authors have often failed to mention the numbers of animals tested, strain, age, or sex of the species studied, or the fume concentration, fume composition, or duration of exposure. This inadequate reporting of methodology is emphasized in relationship to the credibility of the authors' conclusions on the toxicity of welding fumes, or gases, or both in animals.

Inhalation exposures, intratracheal administration of welding fume suspensions, intraperitoneal and subcutaneous injection of the same, and exposure to welding radiation are summarized below. For the interested reader details of these studies are provided in Appendix D.

Inhalation of Welding Fumes and Gases

Experiments on rats, mice, guinea pigs, cats, and rabbits have been reported. In addition, Tables 36 and 37 at the end of this section summarize the acute and chronic effects of this type of exposure.

Rats

No mortality occurred among 10 rats (strain, age, and sex not specified) which were exposed for 20 days to welding fumes from burning 2.4 kg of nickel chromium electrode UONI-13/45 (see Table D1 for fume content). Fume concentration was not specified. In contrast, all eleven rats died by the seventh day of exposure to fumes from 1.4 kg of electrode 606-11 (also described in Table D1) (Ref. 234).

Pulmonary edema and respiratory tract irritation occurred with acute exposures to welding fumes and gases from burning iron-based covered electrodes in two studies (Refs. 235 and 236). In one study (Ref. 235) rats (strain, sex, and age not provided) were exposed to 1600 to 2600 mg/m^3 welding fumes for 6 hours. Autopsies revealed pulmonary edema and respiratory tract irritation. Two rats similarly exposed died of bronchopneumonia. Pulmonary edema, hemorrhage, and peribronchial nodules developed in rats exposed for two half-hour periods daily for 2 months to 1600 to 2600 mg/m^3 of welding fumes (Ref. 235). McCord et al., (Ref. 236), observed multiple minute lung abscesses and siderosis on one albino rat that died (24 were exposed) after 22 days of exposure, 6 hours a day, 5 days a week, to welding fumes and gases containing 444 mg/m^3 iron oxide, 24 ppm nitrogen oxides, and 2 ppm ozone. The author felt this was not a significant finding, however.

Other acute inhalation effects reported in rats included methemoglobinemia and decreased rate of weight gain, noted by McCord et al., (Ref. 236), following exposure of albino rats for 6 hours daily, 5 days weekly, for 43 days to welding fumes and gases described in the previous paragraph. Methemoglobin levels were as high as 15.0 percent in exposed male rats; control values averaged 3.7 percent. Weight gain averaged 32 g in controls, compared to 2.9 g in exposed rats. Harrold et al.,

(Ref. 192), noted slight reductions in rate of weight gain (weight not specified) in albino rats exposed to welding fumes and gases containing 398 mg/m^3 iron oxide, 70 ppm nitrogen oxides, and 32 ppm ozone for 6 hours daily, 5 days weekly, for 38 days.

Pulmonary alveolar macrophages laden with iron oxide particles were noted in autopsies performed after acute exposures in some experiments (Refs. 192, 237, and 238). This occurred after 30 minutes to 4 hours of the exposure of male albino CSE rats to 1500 mg/m^3 welding fumes from rutile covered electrodes (Ref. 237).

Chronic exposure of rats to inhalation of various types of welding fumes has resulted in pulmonary fibrosis in many studies (Refs. 19 through 23 and 239 through 241). Rats exposed to 150 to 180 mg/m^3 of welding fumes from CO_2 shielded arc welding of mild steel with bare electrodes for 4 hours daily, 6 days weekly, for up to 6 months, then autopsied, showed a weak fibrotic lung reaction (Ref. 19). White rats exposed for 3 hours daily, 5 days weekly, for up to 1 year to 290 to 310 mg/m^3 welding fumes from rutile covered electrodes developed increased lung collagen content (see Table D3), indicating fibrosis (Ref. 20). The severity of these changes was proportional to the duration of exposure. Rats (strain, age, and sex not specified) developed pulmonary fibrosis following 1 to 10 months of exposure (daily or weekly exposures not specified) to 50 to 80 mg/m^3 (Ref. 239) or to 120 to 150 mg/m^3 (Ref. 22) welding fumes from K-100 electrodes containing copper oxide. Mongrel albino rats exposed to 70 to 80 mg/m^3 of welding fumes containing copper and nickel oxides also developed fibrosis during the 3 hour daily exposures over 4 to 9 months (Ref. 23). Welding fumes from argon arc welding of aluminum or aluminum-magnesium alloys, 120 to 140 mg/m^3, caused pulmonary fibrosis and increased lung collagen content in white rats exposed 3 hours daily for up to 12 months (Ref. 240). Other lung changes noted in these experiments included pulmonary abscesses (Ref. 240), emphysema (Refs. 21, 23, and 239), hemorrhagic bronchopneumonia (Refs. 24 and 239), and bronchitis (Refs. 21 and 242).

Findings other than lung effects in rats chronically exposed to welding fumes and gases have included gastrointestinal tract inflammation (Ref. 242), reduced fertility (Refs. 243 and 244), liver function alterations (Ref. 245), and central nervous system structural and functional abnormalities (Ref. 19). Chronic gastric and intestinal mucosal inflammation was reported by Lehmann (Ref. 242) in rats (strain, sex, and age not specified) exposed to 34 mg/m^3 fumes from FOX EV 50 electrodes for 6 hours daily for an unspecified duration. Fumes contained iron, calcium, and manganese, plus 120 ppm CO. This exposure also caused bleeding from the nasal membranes and 3 of 18 rats died.

Mature male and female Wistar rats exposed to 222 mg/m^3 of welding fumes from Polish EP 47-28 rutile covered electrodes for 3 hours daily for up to 100 days, then mated with unexposed rats, showed abnormalities of the reproductive system. Females had decreased pregnancy rates, a decreased number of live fetuses per litter, and offspring of low birth weight (89 g in controls, compared to 54 g in exposed females). Females exposed for 80 days and allowed to recover for 102 days prior to mating were infertile. Inorganic iron deposits (siderosis) were detected in cells of the uterus, ovaries, decidua, and placenta (Ref. 243). Male rats, exposed for 100 days and immediately mated with unexposed females, failed to impregnate the females. When exposed for 100 days and allowed to recover for 80 days prior to mating, four of 16 females became pregnant. Edema of the interstitial tissues of the testes, siderosis, degeneration of the germinal epithelium, and an absence of spermatogenesis in seminal ducts were noted on autopsy of males exposed for 102 days to these welding fumes (Ref. 244).

A study of homogenized liver tissue from rats exposed to unspecified concentrations of welding fumes from EP 47-28 covered electrodes for 15 weeks, 3 hours daily, 6 days weekly, and allowed to recover for up to 40 days, revealed no statistically significant difference in oxygen consumption from control values, although the levels obtained were quite variable as shown in Table D6 (Ref. 245).

Central nervous system abnormalities in albino rats exposed to 150 to 180 mg/m^3 fumes and gases from CO_2 shielded arc welding of mild steel (2.1 percent manganese) with bare electrodes for up to 6 months, 4 hours daily, 6 days weekly, were reported by Erman and Rappoport (Ref. 19). Electrophysiological studies revealed significantly increased sensitivity of the caudal neuromuscular apparatus to electrical stimulation (rheobase, chronaxy, and lability indices) as shown in Table D5. Alterations in cholinesterase activity and significant changes in acetylcholine content of the cerebral cortex and medulla oblongata, measured as μg acetylcholine per minute per g of tissue, were noted. Histological examination of the brains of exposed rats revealed cortical and subcortical degenerative changes, the severity of which were directly related to the duration of exposure.

Rabbits

Von Haam and Groom (Ref. 235) found rabbits to be more resistant to welding fume toxic effects than rats, guinea pigs, or mice. Essentially no toxicity, except for one case of bronchial ulceration, occurred in rabbits (sex, strain, and age not specified) exposed for 6 hours to 1600 to 2600 mg/m^3 of welding fumes. Two half-hour exposures daily for 2 months likewise had no toxic effect.

Reduced rates of weight gain were noted in experiments by McCord et al. (Ref. 236) and Harrold et al. (Ref. 192). McCord et al. exposed rabbits (sex, strain, and age not specified) to welding fumes containing 444 mg/m^3 iron oxides, 24 ppm nitrogen oxides, and 2 ppm ozone for 6 hours daily, 5 days weekly, for 45 days. Exposed rabbits gained an average of 1 kg, whereas controls gained 1.5 kg during this same time. Harrold et al. used the same exposure schedule for 38 days to welding fumes containing 398 mg/m^3 iron oxide, 70 ppm nitrogen oxides, and 32 ppm ozone. In both experiments, 90 to 100 percent of all animals survived the exposure without developing lung abnormalities. Possibly, the ozone and nitrogen oxide levels were in error.

Acute respiratory tract irritation and lung damage resulted in one study of rabbits inhaling welding fumes. Titus et al. (Ref. 246) in 1935 exposed rabbits (sex, strain, and age not specified) for 48 to 510 minutes to fumes and gases from dc cutting of mild steel with bare wire electrodes. The fumes contained up to 250 mg/m^3 iron oxide. Tracheal inflammation, pulmonary hemorrhage, and edema occurred; 2 of 16 died. When these fumes were filtered to remove particulates, exposed rabbits also developed pulmonary edema and 2 of 12 died, leading Titus et al. to conclude that the gases (ozone and nitrogen oxides, levels not specified) were the toxic components of the exposure.

Chronic pulmonary effects, described as minimal thickening of lung septae with no significant fibrotic nodulation, were documented by Garnuszewski and Dobrzynski (Ref. 27). They exposed rabbits to welding fumes and gases at concentrations typically found in shipyard welding environments (concentration not specified) for 4 hours daily for 180 days. Fumes contained 23 percent iron oxide, 8 percent silica, 14 percent titanium dioxide, 9 percent manganese dioxide, and less than 1 percent oxides of aluminum, calcium, nickel, and copper.

McCord et al. (Ref. 236) additionally reported methemoglobinemia in rabbits exposed to welding fumes and gases as described above. Unexposed controls averaged 0.6 percent methemoglobin, while exposed rabbits had levels of 2.7 to 2.9 percent methemoglobin. Termination of exposure was accompanied by a return of methemoglobin to control values.

Guinea Pigs

Guinea pigs were most sensitive to acute toxic effects of inhaling welding fumes when compared to the sensitivities of rats, rabbits, and mice (Ref. 247). One hour of exposure to welding fumes (concentration not specified) from basic covered electrodes LB-52 (see Tables D9 and D10 for composition) was lethal only to the guinea pig. In further experiments using identical exposure conditions, 10 of 12 guinea pigs died within 24 hours following fume inhalation. Pulmonary edema, methemoglobinemia, and bronchopneumonia occurred.

Other reports of guinea pigs exposed to welding fumes lead to the conclusion that lung reactions are usually mild or absent (Refs. 235 and 248), although occasional severe pulmonary effects have been documented (Refs. 27, 235, and 247).

Pulmonary edema caused the death of 1 of 3 guinea pigs exposed to 1600 to 2600 mg/m^3 of welding fumes for 6 hours (Ref. 235). Three of 5 guinea pigs exposed to the same fume levels for two half-hour periods daily for 2 months developed severe lung reactions (not further described) and died. Thirty of 72 guinea pigs died of pulmonary edema, or bronchopneumonia, or both during a 110-day exposure, 4 hours daily, to welding fumes at concentrations usually present in the shipyard welding milieu (fume concentration not specified). In animals surviving the total exposure, autopsies revealed fibrotic nodulation, thickened alveolar septae, and siderosis (Ref. 27).

Mice

No abnormalities resulted from the exposure of mice (sex, strain, and age not specified) to 1600 to 2600 mg/m^3 welding fumes for 6 hours. However, the filtered welding fumes, containing nitrogen oxides (level not specified), traces of CO, and 0.7 percent CO_2, but no particulates, caused the death of all 4 mice exposed (duration of exposure not specified). The authors concluded that mice were the least resistant to acute toxicity of welding fumes and gases of all of the species they tested (rats, rabbits, and guinea pigs) (Ref. 235).

In albino mice exposed for 6 months, 4 hours a day, to 150 to 180 mg/m^3 fumes from arc welding using CO_2 as a shielding gas, and containing 3.6 to 6.8 mg/m^3 nitrogen oxides (1.92 to 3.62 ppm based on NO_2), and 0.18 ppm ozone and 122 to 140 ppm CO, the authors noted decreased rate of weight gain, increased kidney and liver weight, depression of cholinesterase activity of the central nervous system, degeneration in the cerebral cortex and subcortex, moderate myocardial dystrophy, and fibrosis in the heart, liver, kidney, and lung stromal tissues.

Cats

Acute exposures of cats (age and strain not specified) to inhalation of welding fumes and gases from dc arc cutting of mild steel containing 60 to 250 mg/m^3 iron oxide for 3 to 6 hours resulted in the deaths of 5 out of 13. Autopsies revealed pulmonary edema and respiratory tract inflammation in all but 3 cats, according to Titus et al. (Ref. 246).

Intratracheal Administration of Welding Fume Suspensions

Effects in rats receiving single intratracheal doses of 20 to 50 mg of suspended or dissolved welding fumes include stimulation of pulmonary phagocytosis (Ref. 20), increased lung content of collagen (Refs. 20 and 24), or hydroxyproline (Refs. 25, 26, 249, and 250), and of ascorbic acid (Ref. 24). Since collagens are the only proteins known to contain a significant amount of hydroxyproline, an increase in the latter is an indication of fibrosis (Ref. 251). These studies are summarized in Table 38 at the end of this section.

In one study (Ref. 25), a 50 mg suspension of welding dust from basic covered electrodes (content not specified) caused the deaths of all of 15 treated male Wistar rats. Increased wet lung weight and phospholipid and hydroxyproline content were also noted. Lung tissues contained fibrotic nodules.

Senczuk and Nater (Ref. 26) reported significantly elevated hydroxyproline and proline levels in lungs of male Wistar rats which were administered 20 mg of welding dust from covered EP50-BNT low fluoride content electrodes (6 percent CaF_2, 5 percent SiO_2, 2 percent Mn in coating). After 84 days, the lung hydroxyproline level was 4.7 mg, compared to 3.2 mg in control lungs. For proline, after 42 days, the levels were 8.6 mg in exposed rats and 4.1 mg in controls. Increased lung weight and reduced rate of weight gain were also noted (see Table D12).

A progressive rise in collagen content was noted in lungs of white rats up to 270 days after intratracheal administration of 50 mg of welding dust suspension containing oxides of manganese, iron, and silica from ANO-1 rutile covered electrodes. Control rats had 2.7 to 3.0 mg of collagen per 100 mg lung tissue; 270 days following the treatment, rats had 4.4 mg of collagen per 100 mg lung tissue. In addition, ascorbic acid elevations were noted in the lungs 270 days after treatment (28.9 mg percent); control values ranged between 20.3 and 22.2 mg percent of ascorbic acid (Ref. 24).

Guskova and Komovnikov (Ref. 20) found intratracheal administration of welding dust increased both alveolar phagocytosis and the phagocytic index in lungs of rats. The exposure was a 50 mg suspension of dust from welding with rutile covered electrodes OZS-4 or TsL-11 (compositions not specified). These two measurements returned to control levels by 12 months after the treatment. In addition, collagen content of the lungs of treated rats increased from 12.4 mg in controls to 25.3 mg in rats 12 months after treatment with welding dust from the OZS-4 electrodes.

Distribution and Excretion of Inhaled Welding Fumes

The distribution of fluorides in lungs of rats and rabbits exposed to fumes from basic covered electrodes containing 17 percent calcium fluoride was studied by Krechniak (Ref. 232). Lung fluoride content increased rapidly for 3 days of exposure (3 hours daily) to 60.6 mg/m^3 of fumes and remained elevated for the total 3 months of exposure and for an additional 20 weeks of recovery after exposures were terminated. After 2 weeks of exposure, the fluoride content of the teeth reached significantly elevated levels in both species. Elevations in basic fluoride content were detected after 30 to 40 days of fume exposure. Bone and tooth fluoride values remained elevated after the end of the exposure period. Blood levels of fluorides remained constant throughout the exposure. Urinary fluoride levels were elevated, compared to the controls; the elevation was maintained from 14 days of exposure until the end of the exposure period (see Tables D15 and D17).

Kukula et al. (Ref. 252) monitored lung, liver, and kidney manganese levels in guinea pigs exposed to welding fumes from EP 47-28P Polish electrodes containing 2.52 mg/m^3 manganese oxides, 4 hours daily for up to 126 days, and allowed to recover for 134 days. Control levels were 0.5 μg/g dry lung weight, 4.1 μg/g dry liver weight, and 1.9 μg/g dry kidney weight. After 126 days of exposure, these levels increased to 494.1, 7.5, and 4.4 μg/g dry weight, respectively. After 134 days of recovery, the levels reported were 145.1, 4.5, and 2.3 μg/g dry tissue, respectively. Lung manganese levels, but not liver or kidney levels, were found to increase with length of exposure and increasing fume concentration. Brain and pancreatic manganese content in guinea pigs exposed for either 6 months or for 126 days, plus 136 days of recovery before being killed, also showed elevations. Only brain levels returned to control values following the 136 days of recovery; pancreatic manganese remained elevated (Ref. 253).

Lung, liver, and kidney silica content was reported in guinea pigs exposed for 4 hours daily for up to 126 days, followed by up to 134 days of recovery, to 4.1 to 4.3 mg/m^3 SiO_2 in fumes from Polish EP 47-28P or EP 52-28P covered electrodes. Lung silica content increased rapidly, from 70 to 90 μg/g dry tissue in controls to 996 μg/g dry tissue by 126 days of exposure, and rapidly decreased during the recovery period to 135 μg/g dry tissue 134 days after exposure. Liver and kidney levels of silica increased and decreased in a similar manner, although the maximum increases were 4 times for the liver levels and 2.5 times for the kidney levels of silica (Ref. 254).

In mixed fume inhalation studies, rats eliminated chromium from lung tissue more slowly than iron, cobalt, or antimony (Refs. 237 and 238). The welding fumes contained 500 mg/m^3 iron, 0.4 mg/m^3 cobalt, 0.1 mg/m^3 chromium, and 1.0 mg/m^3 antimony. The exposures were 30 minutes or 4 hours. No retention of these 4 elements occurred in liver or kidney tissues of animals killed 1 to 10 weeks after the exposure.

Carcinogenicity

Migai and Norkin in 1965 (Ref. 234) intratracheally administered to 10 rats (strain, age, and sex not specified) a suspension of 50 mg of dust from welding with 606/11 electrodes. Fumes contained 20.2 mg/m^3 MnO_2, 22.5 mg/m^3 CrO_3, 32.1 mg/m^3 F, and 4 mg/m^3 SiF_4. Rats were killed 1.5 years later. No evidence of lung tumors could be found. Another 10 rats inhaled these fumes for 9 months and similarly exhibited no tumors in 1.5 years.

In other inhalation studies in this report, lung tumors, as an incidental finding, have not been detected.

Metal Fume Fever

There is an almost total lack of animal investigations into causes of metal fume fever and mechanisms by which temporary immunity is conferred to exposed individuals. In one study (Ref. 247), body temperature and white blood cell counts were monitored in mature male rabbits (strain not specified) exposed to fumes from basic covered LB-52 or LBM-52 electrodes or ilmenite covered B-17 electrodes (fume levels not specified) for up to 1 hour. Average body temperature elevations were 0.04 to 0.13° C over a 10-hour period following exposure to LBM-52 electrodes, and 0.38° C in rabbits inhaling fumes from LB-52 basic covered electrodes. Leukocyte counts increased 10 percent in 3 of the 6 rabbits exposed to LB-52 electrode fumes, and in 1 of 6 rabbits exposed to either LBM-52 or B-17 electrode fumes. No statistical comparison with unexposed controls was presented. The significance of these slight temperature changes and white blood cell elevations cannot be fully evaluated from these data.

The Effect of Welding Fumes on Influenza or Pulmonary Tuberculosis

Naumenko and Frolov (Ref. 255) determined that there was a possible association between the influenza virus and the toxicity of welding fumes. Rats treated by intratracheal administration of a 50 mg suspension of welding fumes (oxides of iron, manganese, silica, and other components at unspecified concentrations) and a 0.2 ml influenza virus suspension (titer 10^{-5}), then killed 40 to 270 days later, showed generally elevated DNA and RNA content in the lungs when compared to rats treated with either welding fumes or the influenza virus alone. However, there was no relationship between DNA or RNA levels and duration of the observation period after treatment.

In white rats, guinea pigs, and mice exposed to welding fumes and infected with tuberculosis, reactivation and spread of lesions from partially healed tubercles did not occur more often than in animals with the tuberculosis infection alone. Lesions in the controls and in the animals inhaling fumes healed at similar rates; progressive tuberculosis did not develop. Larger and more numerous tubercles were noted in animals infected after 10 months of exposure to welding fumes. Fumes were generated from covered mild steel electrodes; high levels of Fe_3O_4 and low Mn, Al, and SiO_2 concentrations were reported (levels not specified). The R_1 attenuated tubercle bacillus strain was diluted to 10 to 15 bacilli per oil immersion field and inhaled in a nebulized suspension (Ref. 256).

Exposure to Arc Welding Radiation

Slight to moderate degrees of injury to intact eyes of experimental animals exposed to welding radiation (infrared, visible, and ultraviolet wavelengths) have been reported. Mice exposed to radiation at a factory welding site (wavelength or intensity not specified) for 2 to 6 hours, then killed, had reduced vitamin B_1 content of liver, brain, and eyes. The author noted that B_1 deficiency causes reduction in the power of accommodation in human eyes (Ref. 161). Walther and Szilagy (Ref. 257) exposed nine rabbits (strain not specified) to arc welding light focused through a 5 diopter quartz lens with a 20 cm focal point for 12 hours. Animals killed immediately thereafter had slight corneal opacification, severe keratoconjunctivitis, and retinal edema. Lactic dehydrogenase activity of the retina was reduced, compared to the unexposed controls' levels; this enzyme change indicated an early stage of retinal damage that was not yet histologically detectable.

In 5 rhesus monkeys exposed to wavelengths from 1064 to 441.6 nm, Ham et al. (Ref. 258) produced thermal burns in the retina with the longer wavelengths, but the shorter wavelength caused photochemical retinal damage. The authors could not distinguish between rod or cone cell damage, nor could they determine the extent of involvement of the retinal photopigments. The photochemical lesions were probably localized in the pigment epithelium. Because the lens and ocular media protectively absorb wavelengths shorter than 440 nm, the number of photons reaching the retina is drastically reduced; retinal damage by these wavelengths would be less likely to occur in the intact eye, although the unprotected retina shows an exponential increase in sensitivity with decreasing wavelengths.

Table 36
Summary of acute inhalation studies in animals

Species	Experimental protocol	Results	Reference
Rats	Up to 20 days of exposure to fumes from burning 2.4 kg of nickel-chromium electrodes (UONI-13/45), (? fume level)	No deaths in 10 animals	Migai and Norkin, 1965 (234)
Male albino CSE rats	30 min. exposure to 1500 mg/m^3 welding fumes from rutile covered electrodes	Brown discoloration of lungs and stomach; lungs contained many alveolar pigmented macrophages	Hewitt and Hicks, 1972 and 1973 (237, 238)
Male albino CSE rats	4 hr. exposure to 1500 mg/m^3 welding fumes from rutile covered	Pigmented macrophages in lung parenchyma	Hewitt and Hicks, 1972 and 1973 (237, 238)
Rats	6 hr. exposure to 1600-2600 mg/m^3 welding fumes from A-5 electrodes*	Pulmonary edema; respiratory tract irritation; death from bronchopneumonia	Von Haam and Groom, 1941 (235)
Rats	Two 1/2 hr. exposures daily for 2 months to 1600-2600 mg/m^3 welding fumes from A-5 electrodes*	Pulmonary edema, hemorrhage; peribronchial nodules	Von Haam and Groom, 1941 (235)
Albino rats	6 hrs. daily, 5 days weekly for 38 days of exposure to welding fumes and gases (398 mg/m^3 Fe_2O_3; 70 ppm nitrogen oxides; 32 ppm ozone)	Slowed rate of weight gain; 90% survived; pulmonary iron deposits	Harrold et al., 1940 (191)
Albino rats	6 hrs. daily, 5 days weekly for 45 days of exposure to welding fumes and gases from rutile covered electrodes (444 mg/m^3 Fe_2O_3; 24 ppm nitrogen oxides; 2 ppm ozone)	Slowed rate of weight gain; methemoglobinemia	McCord et al., 1941 (236)
Rabbits	48-510 min. exposure to 35-250 mg/m^3 fumes and gases from dc cutting of mild steel	Pulmonary edema; respiratory tract inflammation; lung hemorrhage; 2 of 16 died	Titus et al., 1935 (246)
Rabbits	313-510 min. exposure to gases from welding same as above (? level of ozone, ? level of nitrogen oxides)	Pulmonary edema, 2 of 12 died	Titus et al., 1935 (246)
Rabbits	6 hr. exposure to 1600-2600 mg/m^3 welding fumes from A-5* electrodes	Bronchial ulceration	Von Haam and Groom, 1941 (235)
Rabbits	Two 1/2 hr. exposures daily for 2 months to 1600-2600 mg/m^3 welding fumes from A-5* electrodes	No abnormalities	Von Haam and Groom, 1941 (235)
Rabbits	6 hrs. daily, 5 days weekly for up to 38 days of exposure to welding fumes and gases (398 mg/m^3 Fe_2O_3; 70 ppm nitrogen oxides; 32 ppm ozone)	Slowed rate of weight gain; 90% survial; pulmonary and tracheobronchial lymph node iron deposits	Harrold et al., 1940 (192)
Rabbits	6 hrs. daily, 5 days weekly for 45 days of exposure to welding fumes and gases from rutile covered electrodes (444 mg/m^3 Fe_2O_3; 24 ppm nitrogen oxides; 2 ppm ozone)	Slowed rate of weight gain; 100% survived; methemoglobinemia	McCord et al., 1941 (236)

Table 36 (continued)
Summary of acute inhalation studies in animals

Species	Experimental protocol	Results	Reference
Guinea pigs	1 hr. exposure to welding fumes from basic covered electrode LB-52 (? fume level)	10 of 12 died in 24 hrs. pulmonary edema; methemoglobinemia; bronchopneumonia	Kawada et al., 1964 (247)
Guinea pigs	1 hr. exposure to gases from welding with basic covered electrode LB-52 (25-60 ppm nitrogen oxides; 3.8 vol % CO_2; trace CO)	No abnormalities	Kawada et al., 1964 (247)
Guinea pigs	6 hr. exposure to 1600-2600 mg/m^3 welding fumes from A-5* electrodes	Pulmonary edema; respiratory tract irritation	Von Haam and Groom, 1941 (235)
Guinea pigs	Two 1/2 hr. exposures daily for 2 months to 1600-2600 mg/m^3 welding fumes from A-5* electrodes	3 of 5 died of severe lung reactions	Von Haam and Groom, 1941 (235)
Mice	6 hr. exposure to 1600-2600 mg/m^3 welding fumes from A-5* electrodes	No abnormalities	Von Haam and Groom, 1941 (235)
Mice	Exposure to gases (? duration) from welding with A-5* electrodes (0.7% CO_2, trace CO, qualitative presence of nitrogen oxides)	4 of 4 died	Von Haam and Groom, 1941 (235)
Cats	3-6 hours exposure to 60-250 mg/m^3 fumes and gases from dc cutting of mild steel	5 of 13 died; pulmonary edema; inflammation of respiratory tract	Titus et al., 1935 (246)

*A-5 electrodes: 17.6% Fe_2O_3, 9.5% Cr_2O_3, 10.6% Mn_3O_4, 16.8% CaO, 16.1% Na_2O, 5% Al_2O_3, 14.6% F

Table 37
Summary of chronic inhalation studies in animals

Species	Experimental protocol	Results	Reference
Rats	6 hr. daily exposures (duration ?) to 34 mg/m^3 fumes from 6-7 kg of FOX EV 50 electrodes	3 of 18 died; inflammation of gastric and intestinal mucous membranes	Lehmann, 1956 (242)
White rats	1-12 months (? daily ? weekly) exposure to 40-60 mg/m^3 fumes, electrodes containing CuO, MnO_2, Fe_2O_3, TiO_2, NiO, V_2O_5	Emphysema; bronchitis; fibrosis; desquamation in airways	Samoilova and Kireev, 1975 (21)
Albino rats	4 hrs. daily, 6 days weekly for 3-6 months exposure to 150-180 mg/m^3 fumes from CO_2 shielded iron welding	Reduced rate of weight gain; increased liver and kidney weight; congested brain, liver, lungs, lungs, kidneys; degeneration in brain, heart, liver, kidneys, ovaries; slight lung fibrosis	Erman an Rappoport, 1970 (19)
Rats	6 hr. daily exposures (duration ?) to 300 mg/m^3 fumes from 6-7 kg of FOX EV 50 electrodes	15 of 18 died of severe inflammatory bronchitis; marked iron deposits in lungs	Lehmann, 1956 (242)
White rats	3 hrs. daily, 5 days weekly for up to 1 year exposure to 290-310 mg/m^3 fumes from rutile covered electrodes (OZS-4 and TsL-11)	Stimulation of phagocytosis in lungs; increase in lung collagen content; slight fibrosis	Guskova and Komovnikov, 1974 (20)
Mature Wistar rats	3-5 months (240-576 hr.) exposure to welding fumes and gases from rutile and basic covered electrodes at a shipyard (? fume level)	Iron deposits in lungs; inflammation and irritation of lungs	Felczak, 1967 (300)
Rats (150 g)	5-6 hrs. daily, 6 days weekly for 9 months of exposure to fumes (? fume level) and gases from nickel-chromium electrodes (606/11)	Oxygen uptake decreased; reduced rate of weight gain; decreased skin sensitivity to electrical stimulation	Migai and Norkin, 1965 (234)
Albino rats	1-10 months (? daily ? weekly) exposure to 50-80 mg/m^3 welding fumes from K-100 electrodes containing copper oxide	25% mortality in 6 months; hemorrhagic pneumonia or bronchopneumonia; focal emphysema; diffuse fibrosis; lymphoid follicular hyperplasia	Vorontsova et al., 1969 (239)
Mongrel albino rats	3 hrs. daily for 4-9 months exposure to 70-80 mg/m^3 welding fumes containing copper and nickel oxides	Fibrosis; emphysema; pulmonary alveolar proteinosis	Arutyunov et al., 1976 (23)
Rats	1-10 months (? daily ? weekly) exposure to 120-150 mg/m^3 welding fumes from K-100 electrodes, containing copper oxide	Pneumonia; fibrosis; increased respiration rate; increased dry lung weight; increased lung collagen content	Vlasova - Pryadilova, 1971 (22)

Table 37 (continued)
Summary of chronic inhalation studies in animals

Species	Experimental protocol	Results	Reference
White rats	3 hrs. daily for up to 12 months of exposure to 120-140 mg/m^3 fumes from argon arc welding with aluminum or aluminum-magnesium electrodes	Pneumonia; bronchitis; pulmonary absecesses; fibrosis; increased collagen content of lungs	Leonicheva, 1965 (240)
Rabbits	4 hrs. daily for 6 months of exposure to welding fumes (? fume level) from Polish electrode EP 47-28P	Thickening of lung septae; no fibrosis	Garnuszewski and Dobrynski, 1966 (27)
Guinea pigs	4 hrs. daily for 6 months of exposure to welding fumes (? fume level) from Polish EP 47-28P	Interstitial pneumonia; siderosis and silicosis; fibrosis	Granuszewski and Dobrynski, 1966 (27)
Guinea pigs	180 days (? daily ? weekly) exposure welding fumes from electrode EP 47-28P (2.52 mg/m^3 Mn; 6.08 mg/m^3 silica; 24.8 mg/m^3 iron) (? fume level)	No abnormalities	Gadzikiewicz and Dominiczak, 1968 (248)
Albino mice	4 hrs. daily for 6 months exposure to 150-180 mg/m^3 fumes and gases from CO_2-shielded arc welding	Decreased rate of weight gain; increase in liver and kidney weight; degeneration of cerebral cortical neurons; myocardial dystrophy; slight pulmonary fibrosis	Erman and Rappoport, 1970 (19)
Albino rats	3 hrs. daily, 6 days weekly for 15 weeks exposure to welding fumes (? fume level) from EP-47-29 Polish rutile covered electrodes	Wide variation in oxygen consumption of liver homogenates compared with controls	Byczkowski et al., 1965 (245)
3-4 month female Wistar rats	32-102 days, 3 hrs. daily exposure to 222 mg/m^3 welding fumes from EP 47-28 Polish rutile covered electrodes, followed by mating with unexposed males	Decrease in numbers of pregnancies, small litter size, low fetal weight, siderotic changes in ovaries, uterus, placenta	Dabrowski et al., 1966 (243)
Mature male Wistar rats	100 days, 3 hrs. daily exposure to 222 mg/m^3 welding fumes from EP 47-28 Polish rutile covered electrodes, followed by mating with unexposed females	Reduced fertility; siderosis in interstitial tissue of testis; degeneration of germinal epithelium; absence of spermatogensis	Dabrowski et al., 1966 (244)

Table 38
Summary of intratracheal effects of welding fume suspensions in animals

Species	Experimental protocol	Results	Reference
150 g male Wistar rats	Single dose of 20 mg of suspension of welding dust from EP50-BNT low fluoride covered electrodes	Progressive increase in lung hydroxyproline and proline content over 1-3 months; reduced rate of weight gain	Senczuk and Nater, 1970 (251)
150 g white male Wistar rats	Single dose at 20 mg of suspension of welding dust from EP 50-BNT low fluoride covered electrodes	Elevated serum amino acid levels 3 wk. after dosing; elevated neuraminic acid level in serum 5-10 days after dosing	Senczuk et al., 1970 (249, 250)
200 g male Wistar rats	Single dose of 50 mg of suspension of welding dust from basic covered electrodes	15 of 15 died in 24 hrs.; increased wet lung weights; increased lung content of phospholipids and hydroxyproline	Kysela et al., 1973 (25)
White rats	Single dose of 50 mg of suspension of welding dust from ANO-1 rutile covered electrodes	Progressive increase in lung collagen and ascorbic acid content over 1-6 months	Naumenko, 1966 (24)
White rats	Single dose of 50 mg of suspension of welding dust from OZS-4 or TsL-11 rutile covered electrodes	Progressive increase in lung collagen content over 3-12 months; stimulation of lung phagocytosis; low fibrogenic potential	Guskova and Komovnikov, 1974 (20)

Chapter 4

Special Studies on Mutagenicity of Welding Fumes

The mutagenicity of welding fumes has been examined using bacterial systems. Results in the Ames Salmonella/microsome mutagenicity tests have been well correlated with carcinogenicity for organic compounds, McCann et al. (Ref. 77); however, for inorganic materials, the relationship has not been well documented.

Preliminary screening studies by Hedenstedt et al. (Ref. 88), using three in vitro mutagenicity assays, indicated that fumes generated in the welding of stainless steel, but not mild steel, was mutagenic. Then, Maxild et al. (Ref. 83) studied the mutagenicity as measured by the Ames test of fumes from shielded metal arc and gas metal arc welding of stainless steel and mild steel. In this test a positive response is produced by reversion of a mutant to the wild type. Fume samples were collected on filter paper and suspended in dimethylsulfoxide. Then 200 μl of suspension were added to pour plates containing *Salmonella typhimurium* tester strains TA 100 (detects base-pair substitution) and TA 98 (detects frame-shift mutagens), both with and without liver homogenate from phenobarbital-treated rats. Fumes from the following were assayed:

1. Shielded metal arc welding using a rutile type covered electrode for welding austenitic stainless steel (18 percent Cr, 10 percent Ni).
2. Shielded metal arc welding using a rutile type covered electrode for acid proof stainless steel (18.5 percent Cr, 12.5 percent Ni, 3 percent Mo)
3. Gas metal arc (argon) welding using an electrode for austenitic stainless steel (18 percent Cr, 8 percent Ni)
4. Gas metal arc (argon) welding using an electrode for high corrosion-resistant stainless steel (18 percent Cr, 8 percent Ni, 2.5 percent Mo)
5. Shielded metal arc welding using a rutile type electrode for mild steel (typically, 1 percent Si, 1 percent C, 1 percent Mn)
6. Gas metal arc (argon-CO_2) welding using an electrode for mild steel

As measured by this test, fumes from welding stainless steel were mutagenic in all cases (fumes 1 to 3), except gas metal arc welding of high corrosion-resistant stainless steel (fume 4); for the mutagenic fumes, increasing fume concentration increased the number of revertants, whether rat liver homogenate was present or not. A concentration of gas metal arc welding fume particles (fume 3) 4 to 8 times higher than the shielded metal arc welding fumes (fume 1 or 2) was required to double the number of revertants (above spontaneous background), indicating that the latter was the more potent mutagen. The stainless steel fume composition is too complex to attribute the mutagenicity to any particular element. However, hexavalent chromium ($CaCrO_4$, K_2CrO_4, Na_2CrO_4, CrO_3), but not trivalent chromium ($CrK[SO_4]_2$, $CrCl_3$), has been independently found to be mutagenic in the Ames test (Ref. 259).

Neither fume from welding mild steel (fume 5 or 6) was mutagenic in this test.

Appendix A
Details of the Exposure

Particulates, gases, radiation from the arc, heat, and noise are the most prominent byproducts of many welding processes. Examples of important members of these classes are enumerated in Table A1. The health hazard potential of any of these individual agents or combination of agents depends on the level of exposure.

Table A1
Chemical and physical agents commonly produced as byproducts of welding operations

Fumes*	Gases	Radiant Energy
Aluminum	Carbon dioxide	Ultraviolet
Chromium	Carbon monoxide	Visible
Copper	Nitrogen oxides	Infrared
Fluorides	Ozone	
Iron		**Other Agents**
Lead		
Magnesium		Heat
Manganese		Noise
Nickel		
Silica, silicates		
Titanium		
Vanadium		
Zinc		

*Including the metals and their oxides

Thus, the hazard potential will be affected by the degree of confinement of the welder in the workplace, by his position while welding, by the duration of the exposure, and by the control measures such as ventilation and personal protective equipment employed for hygiene and safety. Furthermore, the composition and quantities of fumes and gases, the spectral variation and intensity of radiation, as well as the amount of heat and noise generated, will depend on the material being welded and welding process employed. Since some fume and gas components, which may be present in some operations, are inherently more toxic than others, both process and filler metal will also affect hazard potential.

The American Welding Society has identified over 80 different types of welding and allied processes in commercial use (Ref. 79). Of these processes, some of the more common types include shielded metal arc welding, gas metal arc welding (SMAW), gas tungsten arc welding (GTAW), submerged arc welding (SAW), plasma arc welding (PAW), flux cored arc welding (FCAW), and oxygas welding (Refs. 82 and 260). Each method has its own particular metallurgical and operational advantages, and each may present its own potential health and safety hazards. This discussion will deal primarily with these most widely employed arc welding processes and the potential hazards associated with them.

Formation of Fumes

Fumes arise principally from the filler metal of consumable electrodes and any covering or fluxing materials they may contain. Vaporization, some reaction with atmospheric oxygen, and then rapid condensation appear to be the main processes in their formation. The base metal weld pool is much cooler than the electrode tip; therefore, the base metal is a significant contributor to total fume only when it contains a volatile protective coating or volatile alloying elements (Ref. 70). Filler metals usually have a composition similar to the base metals to be welded. When the filler metal is the only major source of fume, it is possible to estimate fume composition. However, the total rate of fume generation cannot be predicted easily since it varies with the type of welding process and depends upon several factors, including the following:

(1) Welding current
(2) Arc voltage
(3) Polarity of the electrode
(4) Electrode diameter
(5) Speed of welding
(6) Welding practices

These factors will be discussed in detail under individual processes and the fumes generated from them.

In the shielded metal arc process, welding (fusion) is accomplished by heating with an electric arc struck between the metals to be joined and the electrode (filler metal). The electrode contains a covering that serves several purposes, including the provision of a shielding gas to prevent oxidation of the molten metal (Refs. 82 and 260). This method is commonly used for welding carbon steels and low alloy steels in most structural applications (Ref. 82). Factors affecting fume composition and generation rate are discussed individually.

Composition

The major source of fumes during shielded metal arc welding is the metallic part of the electrode and its covering. Since the composition of filler and base metals are similar, the contribution from the metallic component of the electrode will depend on the metal being welded.

The composition of the electrode covering largely determines the performance of an electrode and the soundness of the weld (Ref. 261). The covering formulations are proprietary, but they contain many classes of chemicals, such as gas generators, slag producers and deoxidizers, alloying substances, and binders (Refs. 80, 131, and 260 through 262). Typical functions and composition ranges of constituents in coatings of mild steel electrodes used for the shielded metal arc process are given in Table A2.

Depending upon the composition of the shielded metal arc electrode coating, some are called acid type electrodes (E6010, E6011), because they contain high percentages of cellulose, or rutile type electrodes (E6012, E6013, E7024), containing significant quantities of TiO_2. Basic (also alkaline or low hydrogen) types (E7018) contain substantial quantities of inorganic carbonates and fluorspar; and neutral types contain low levels of titania, carbonate, cellulose, and fluorspar (Ref. 261).

The effect of the type of coating on the fume composition is presented in Tables A3 and A4. Table A3

Table A2
Functions and composition ranges of constituents of coatings for mild steel electrodes

Constituents of coating	Function of constituent	Composition range,%, in coating of class of electrodes		
		E6010, E6011	E6012, E6013	E7018
Cellulose	Shielding gas	25-40	2-12	-
Calcium carbonate	Shielding gas, fluxing agent	-	0-5	15-30
Fluorspar	Slag former, fluxing agent	-	-	15-30
Rutile (titanium dioxide)	Slag former, arc stabilizer	10-20	30-55	0-5
Feldspar	Slag former	-	0-20	0-5
Mica	Extrusion, arc stabilizer	-	0-15	-
Clay	Extrusion, arc stabilizer	-	0-10	-
Asbestos	Slag former, extrusion	10-20	-	-
Iron powder	Deposition rate	-	-	25-40
Ferrosilicon	Deoxidizer	-	-	5-10
Ferromanganese	Alloying, deoxidizer	5-10	5-10	2-6
Sodium silicate	Binder, fluxing agent	20-30	5-10	0-5
Potassium silicate	Arc stabilizer, binder	-	5-15	5-10

American Society of Metals, 1971 (Ref. 261).

Table A3
Composition of fumes produced during shielded metal arc welding of steel with bare and coated electrodes

Electrode type	Fume content, wt %					
	Fe_3O_4	SiO_2	MnO	CaO+MgO	CaF_2	Al_2O_3
Bare electrode	92.45	2.11	4.95	0.38	---	---
Basic electrode	34.85	7.70	5.34	3.48	27.3	0.93
Acid electrode	55.97	28.20	10.07	1.55	---	2.00

Hummitzsch, 1955 (Ref. 263).

Table A4
Analysis of fumes produced during shielded metal arc welding of mild steel with coated electrodes

Electrode type	Arc current, A	Arc voltage, V	Total fumes, g/electrode	Fume analysis, wt %					
				Fe_2O_3	SiO_2	TiO_2	MnO_2	CaO	F^-
Neutral	200	32	0.5-1.5	36.2	31.4	<0.1	14.4	---	---
Acid	135	29	<1.0	26.6	45.8	1.6	12.2	---	---
Rutile	160	21	<0.5	70.9	10.7	3.3	7.1	---	---
Basic (lime-fluoride)	210	25	1.0-2.5	24.8	6.7	<0.1	4.7	16.6	20.7

Thrysin et al., 1952 (Ref. 264).

presents the work of Hummitzsch, published in Germany in 1955 (Ref. 263). It illustrates the contribution to fume of the basic and acidic type coatings as opposed to bare wire. Silicates (and possibly silica) are important fume constituents when a coating is present. Table A4 presents the variations in composition of fumes from Swedish electrodes available in 1952, when mild steel (0.8 Mn, 0.5 Si, 0.25 C) was welded with neutral, acid, rutile, and basic coated electrodes (Ref. 264). The presence of manganese from the neutral and acid coatings is notable. Also, in both tables the presence of fluoride is prominent when basic electrodes are used.

Historically, from the analyses of fumes in several shipyards by Dreessen et al., 1947 (Ref. 3), it has been indicated that 50 percent of the total fume content (by weight) was iron oxide, 15 percent titanium oxide, 8 percent silicon oxide, and the remainder was a mixture of acid soluble metals such as magnesium, calcium, aluminum, manganese, chromium, and copper. In another investigation by Tebbens and Drinker, 1944 (Ref. 265), designed to study the fumes evolved during welding in shipyards with electrodes (E6010, E6011, E6012) available then, it was reported that the average fume sample contained 60 to 70 percent iron oxide and titanium oxide, 10 to 20 percent silica, and 2 to 12 percent manganese oxide.

Recently, Patteè et al., 1978 (Ref. 70), completed a study of fumes for the American Welding Society using electrodes currently available in the U.S. They analyzed fume compositions generated during the welding of carbon and low alloy steels, as well as stainless steel and high alloys, for selected elements using 3.97 mm diameter electrodes in all cases. The results of this analysis are presented in Table A5. Note the levels of chromium, nickel, and manganese in these fumes.

Welding of aluminum alloys using the shielded metal

Table A5
Concentrations of selected elements in the fumes produced by shielded metal arc electrodes

Electrode	Fume sample weight, g	Concentration, weight %									
		Fe	Mn	Si	Ni	Cu	Cr	Mo	Al	Mg	F
Carbon and low alloy steel											
E6010	0.49	47.5	3.0	5.7	---	---	---	---	---	---	---
E6013	0.16	61.4	5.1	12.2	---	---	---	---	---	---	---
E7018	0.26	26.1	4.5	<0.2	---	---	---	---	---	---	13.1
E7024	0.30	30.2	5.3	18.3	---	---	---	---	---	---	---
E8018C3	0.12	45.2	7.2	---	0.3	---	0.1	<0.1	---	---	35.8
E9018B3	0.16	21.9	5.9	---	0.1	---	1.6	<0.1	---	---	28.1
Stainless steel and high alloy											
E316-15	0.15	8.4	7.7	---	1.1	---	5.8	<0.1	---	---	---
E316-16	0.17	10.0	8.8	---	1.5	---	6.5	<0.1	---	---	17.2
E410-16	0.11	33.1	5.2	---	<0.1	---	---	---	---	---	---
ENi-CI	0.35	2.5	0.3	---	6.9	<0.1	---	---	---	---	10.0
ENiCu-2	0.31	0.1	2.1	---	4.2	6.2	---	---	---	---	---
Inconel 625	0.25	0.6	---	---	4.6	0.7	5.9	2.1	---	---	---
Haynes C-276	0.35	0.3	0.3	---	1.1	---	2.5	0.6	1.0	1.4	5.9
Haynes 25	0.25	---	4.6	---	1.8	---	6.9	---	1.1	0.1	7.7

Pattee et al., 1978 (Ref. 70).

Table A6
Rate of fume generation for various covered electrodes used in shielded metal arc welding*

Electrode type	Current range, A	Fume generation rate, g/min	Weight of fumes-weight of deposited metal, g/kg
Carbon and low alloy steel			
E6010	140-150	0.83	35.85
E6013	145-160	0.31-0.58	14.16-25.75
E7018	170-180	0.57-0.60	20.35-21.83
E7024	200-230	0.43-0.55	8.92-11.11
E8018C3	160-175	0.43-0.47	15.92-17.82
E9018B3	160-180	0.36-0.46	11.19-14.94
Stainless steel and high alloy			
E316-15	150-155	0.28-0.38	8.02-11.08
E316-16	145-150	0.21-0.31	6.56-11.92
E410-16	145-160	0.28-0.34	11.75-13.97
ENi-CI	135	0.37	12.90
ENiCu-2	145	0.31	10.08
Inconel 625	140-155	0.32	9.24
Haynes C-276	130-135	0.37	14.20
Haynes 25	135-140	0.26	8.94

*All electrodes were 5/32 in. (3.97 mm) in diameter.

Pattee et al., 1978 (Ref. 70).

arc process produces aluminum oxide fumes in considerable concentrations. Similarly, when copper alloys (Cu-Zn, Cu-Ni, and Cu-Zn-Sn-Ni) are welded with coated bronze and monel (Ni-Cu) electrodes, the fumes contain copper oxide, zinc oxide, and tin oxides along with fluorides, since the electrodes used in the process are of the basic (lime-fluoride coated) type (Ref. 82).

Fume Generation Rate

Pattee et al. (Ref. 70) also studied various covered electrodes used in welding carbon and low alloy steels and stainless steels from the point of view of fume generation rates. The currents employed were in the middle-to-upper portion of the range recommended by the electrode manufacturer. The rates of fume generation by various electrodes are given in Table A6. These ranged from 0.21 to 0.83 g/min. More fumes were produced from carbon steel electrodes than from stainless steel or high alloy steel electrodes within the same range of welding currents. This may be due to the difference in the composition of the electrode coverings. Fume generation rates and ratios of weight of fume-to-weight of deposited metal were relatively higher for E6010 electrodes than for other covered electrodes. The fumes generation rates of the E7024 (rutile type) electrodes were comparable to those of other carbon and low alloy steel electrodes, such as E8018C3 and E9018B3, despite the fact that the covering on E7024 electrodes was much thicker than the coverings on the other electrodes. The ratio of weight of fume-to-weight of deposited metal for E7024 electrodes was lower than those of other carbon and mild steel electrodes (Ref. 70).

The rate of fume generation increases with increasing welding current (Refs. 70, 81, and 266). Kobayashi et al. (Ref. 266) showed that the fume generation rate (mg/min) was proportional to the current raised to a power, the exponent varying between 1.17 to 1.73, depending upon the composition of the electrode covering. Pattee et al. (Ref. 70) found similar results; the exponent was 2.24 for E6010 and 1.54 for E7018 electrodes.

Arc voltage and arc length are interdependent. Increasing either (at constant or normalized current) increased the fume generation rate (Ref. 70). This was also the conclusion of Kobayashi et al. (Ref. 266) who also showed that electrode polarity affected the fume

Table A7
Typical flux and slag compositions for three types of carbon dioxide shielded flux cored electrodes

Compound or element	E70T-1 or E70T-2, high titania (Nonbasic)		E70T-1, lime-titania (neutral or basic)		E70T-1 or E70T-5, lime (basic)	
	Flux	Slag	Flux	Slag	Flux	Slag
SiO_2	21.0%	16.8%	17.8%	16.1%	7.5%	14.8%
Al_2O_3	2.1	4.2	4.3	4.8	0.5	---
TiO_2	40.5	50.0	9.8	10.8	---	---
ZrO_2	---	---	6.2	6.7	---	---
CaO	0.7	---	9.7	10.0	3.2	11.3
Na_2O	1.6	2.8	1.9	---	---	---
K_2O	1.4	---	1.5	2.7	0.5	---
C	0.6	---	0.3	---	1.1	---
Fe	20.1	---	24.7	---	55.0	---
Mn	15.8	---	13.0	---	7.2	---
CaF_2	---	---	18.0	24.0	20.5	43.5
MnO	---	21.3	---	22.8	---	20.4
Fe_2O_3	---	5.7	---	2.5	---	10.3
Flux	13	---	13	---	27	---

American Society of Metals, 1971 (Ref. 261)

generation rate and that under similar welding conditions, a small diameter electrode generated more fumes than a large diameter electrode (Ref. 266).

The work practices of welders also contribute to some extent to the amount of fumes generated during welding operations. An inclination of the electrode in the forward direction of welding increased the amount of fumes, and the amount of an inclination of 45° increased the fumes by 35 percent, as compared to welding at an angle of 90°. A 20° inclination produced almost twice the amount of fumes as an angle of 65° (Refs. 70 and 266).

Flux Cored Arc Welding

In flux cored arc welding, fusion is produced by heating with an arc between the base metal and a consumable electrode. These electrodes have a central core that contains ingredients that act as deoxidizers, fluxing agents, slag formers, and shielding agents. Additional shielding gas, usually externally supplied carbon dioxide, may also be used (Refs. 82 and 260). The process is used for welding carbon steels, low alloy steels, and stainless steels.

Composition

The filler metals of electrodes used for flux cored arc welding are of two types: one used for gas shielded welding and the other for self-shielded welding. Typical flux and slag compositions of gas shielded flux cored electrodes are given in Tables A7 and A8, respectively.

Pattee et al. (Ref. 70) measured the composition of fumes from welding carbon, low alloy, and stainless steels with electrodes currently used in the U.S. Their results are summarized in Table A9. Since the electrodes are similar in composition to shielded metal arc electrodes, the composition of the fumes generated by these two processes should be similar.

Fume Generation Rate

Heile and Hill (Ref. 81) determined the rate of fume formation during various arc welding processes. Flux cored arc welding produced maximum amounts of fumes as compared to the shielded metal arc and the gas shielded processes. Three types of flux cored electrodes were utilized: E70T-5 silica-base CO_2 shielded; E70T-1 rutile-base CO_2 shielded; and E70T-4 fluorspar-base self-shielded. The rate of fume generation from these electrodes is presented in Table A10.

There was a considerable difference in the rate of fume generation between self-shielded E70T-4 and E70T-1 carbon dioxide shielded electrodes. Self-shielded electrodes produced more fumes than CO_2 shielded electrodes. These differences were due to variations in arc stability and the calcium fluoride content of the electrodes. A similar difference was observed between car-

Table A8
Typical flux and slag compositions for four types of self shielding flux cored electrodes

Compound or element	E70T-4, E60T-7, fluorspar-aluminum		E70T-3, fluorspar-titania		E70T-6, fluorspar-lime-titania		E70T-5, fluorspar-lime	
	Flux	Slag	Flux	Slag	Flux	Slag	Flux	Slag
SiO2	0.5%	---	3.6%	0.2%	4.2%	1.8%	6.9%	0.2%
Al	15.4	---	1.9	---	1.4	---	---	---
Al_2O_3	---	11.8%	---	6.5	---	6.0	0.6	12.8
TiO_2	---	---	20.6	27.0	14.7	33.5	1.2	2.3
CaO	---	---	---	---	4.0	---	3.2	4.1
MgO	12.6	9.2	4.5	4.5	2.2	6.0	---	0.9
K_2O	0.4	---	0.6	1.8	---	---	---	---
Na_2O	0.2	---	0.1	1.0	---	---	0.6	0.9
C	1.2	---	0.6	---	0.6	---	0.3	---
Fe	4.0	---	50.0	---	50.5	---	58.0	---
Mn	3.0	---	4.5	---	2.0	---	7.9	---
CaF_2	63.5	76.1	22.0	53.0	15.3	47.5	22.0	73.7
Fe_2O_3	---	2.5	---	1.9	---	3.6	---	3.0
Flux	18	---	18	---	26	---	26	---

American Society of Metals, 1971 (Ref. 261).

Table A9
Composition ranges of some common elements in fumes from welding with several flux cored electrodes

	Composition range					
Electrode	Fe	Mn	Si	Ni	Cr	F
E70T-1[a,b]	25.2-41.2	6.2-13.5	1.0-7.5	---	---	0.06-6.3
E70T-4[c,d]	11.5-15.1	1.0-3.3	<0.05	<0.01	---	1.04-2.73
E70T-5[b,e]	26.7-29.2	10.9-11.3	<0.05-0.09	---	---	2.63-4.80
316L[c]	12.4	7.3	0.05	1.06	12.5	11.5

Pattee et al., 1978 (Ref. 70).

a. Five samples
b. CO_2 shielded
c. Self-shielded
d. Three samples
e. Two samples

Table A10
Rates of fume generation from various flux cored arc welding electrodes

Electrode	Current, A	Voltage, V	Shielding gas	Fume generation rate, g/min	Weight of fume-weight of deposited metal, mg/g
E70T-1, rutile-base	300	28	CO_2	0.75	11
	350	30		0.96	12
	400	31		1.10	12
	450	32		1.20	11
E70T-5, silica-base	300	28	CO_2	1.40	18
	350	29		1.90	20
	400	30		2.10	21
	450	31		2.50	22
E70T-4, fluorspar-base	250	29	Self-shielded	1.08	18
	325	30		1.62	20
	400	31		2.00	20
	475			2.50	20

Heile and Hill, 1975 (Ref. 81).

bon dioxide shielded E70T-1 and E70T-5 electrodes, where the latter produced more fumes under identical current and voltage conditions. The fume quantities from E70T-5 and E70T-4 electrodes were comparable. In contrast to the shielded metal arc process, the amount of fume produced in the flux cored arc process did not appear to be dependent upon arc voltage.

Similar trends in the rates of fume generation were observed in the study by Pattee et al. (Ref. 70). Flux cored electrodes examined in this study were used to weld carbon, low alloy, and stainless steels. The results of this study are shown in Table A11. Fume generation rates were dependent upon the welding current, increasing as the current was raised. Since the currents used with flux cored electrodes are higher than those used in welding with covered electrodes, more fumes were produced during flux cored welding than during shielded metal arc welding. The highest fume generation rates were encountered with CO_2 shielded E70T-5 electrodes. The E70T-5 electrodes contained appreciable amounts of fluorides and their presence enhanced the production of fumes (Ref. 70).

The fume generation rates and ratios of weight of fumes-to-weight of deposited metal for low alloy steel electrodes (81-C3 and 91-B3) were low when compared to rates and ratios for carbon steel flux cored electrodes (Ref. 70).

The effects of type of shielding gas on flux cored welding were also examined. The fume generation rate was found to be higher when carbon dioxide was used as a shielding gas than when an argon-carbon dioxide mixture was used (Ref. 70).

Table A11
Rates of fume generation of
2.38 mm diameter flux cored electrodes

Electrode	Current, A	Shielding gas	Fume generation rate, g/min	Weight of fume-weight of deposited metal g/kg
Carbon steel				
E70T-1	435-485	CO_2	0.96-2.27	6.65-17.51
E70T-4	370-390	Self-shielded	1.86-2.09	12.76-13.83
E70T-5	425-450	CO_2	2.26-3.25	17.87-23.63
Low alloy steel				
81-C3	440-445	CO_2	1.11	8.69
91-B3	450	CO_2	1.15	8.42
Stainless steel				
E308LT-3	440-445	Self-shielded	1.64	9.11
E316LT-3	340-405	Self-shielded	1.34-2.48	6.97-12.32

Pattee et al., 1978 (Ref. 70).

Gas Metal Arc Welding

In the gas metal arc process, an arc is struck between the base metal and a continuously supplied consumable electrode, which provides filler metal for the weld. The electrode is bare; it contains no coating or core. Shielding of the weld is supplied by an external gas, usually argon, argon-oxygen, argon-carbon dioxide, helium, or carbon dioxide, or mixtures of inert and oxidizing gases. Fumes originate only from the filler metal of the electrode and weld metal pool.

This process is widely used to weld almost all types of weldable metals, such as steel, copper, aluminum, and other alloys (Refs. 80 and 82). The content of the fumes evolved during welding greatly depends upon the compositions of the electrode and the base metal. Other variables to consider are the volatility of the metal constituents, the transfer mode and arc temperature, stability of the arc, and the oxidizing potential of the shielding gases (Ref. 81).

Composition

The composition of fumes should be easier to predict for this process, since the only source is the metal from the electrode and the base metal. When aluminum (1100 Al) was welded using a filler wire of the same composition, the major constituent of the fume was aluminum oxide (99.9 percent). When the same base metal was welded using 4043 Al filler metal (an alloy containing 5 percent silicon), the welding fumes contained 88.5 percent aluminum oxide and 11.5 percent silicon oxide (Ref. 267).

Pattee et al. (Ref. 70), determined the percentage of selected metals in the fume for various electrodes in use in the U.S. today. For reference, these are presented in Table A12.

The composition of the fumes varied considerably with the transfer mode and the oxidation potential of the shielding gas during gas metal arc welding (Ref. 70). Thus, iron oxide content was highest when welding was performed in the spray transfer mode with either Ar-2% O_2 or Ar-9% CO_2 shielding and lowest during welding in the globular transfer mode with CO_2 shielding. In spray transfer, the diameter of the drops of molten metal is either smaller than or equal to the diameter of the electrode when transferred to the base metal. In globular transfer, the drops of molten metal from the electrode have a greater diameter on the average than those deposited in the spray transfer range. In both of these transfer modes, the electrode does not come in contact with the base metal. When the transfer takes place with the electrode in contact with the base metal, the mode is called short circuiting transfer (Ref. 2). Manganese contents were highest when welding was done in either the spray transfer mode with Ar-9% CO_2 shielding or in the globular transfer mode with CO_2 shielding. Lowest manganese contents were observed when welding in the spray transfer mode with Ar-2% O_2 shielding.

Silica contents of the fumes were highest during welding in the globular transfer mode with CO_2 shielding.

Fume Generation Rate

Pattee et al. (Ref. 70) also recently compared the fume generation rates and ratios of weight of fumes-to-

Table A12
Concentrations of selected constituents produced by gas metal arc welding electrodes

Electrode	Shielding gas	Fume sample, g	Concentration, weight %								
			Fe	Mn	Si	Ni	Cu	Cr	Mo	Al	Mg
E70S-3	CO_2	0.40	55.4	5.5	2.5	---	1.2	---	---	---	---
	$Ar+9CO_2$	0.54	62.0	4.6	1.5	---	0.99	---	---	---	---
	$Ar+2O_2$	0.50	63.6	4.8	1.7	---	1.08	---	---	---	---
E70S-5	N.S.	0.11	61.7	5.8	0.93	---	1.75	---	---	---	---
ER5356	N.S.	1.06	---	---	---	---	---	---	---	38.0	3.8
ER4043	N.S.	0.10	---	---	1.7	---	---	---	---	46.2	---
ERNiCu-7	N.S.	0.13	5.0	1.1	0.65	22.1	44.4	<0.01	---	---	---
Inconel 625	N.S.	0.07	1.9	---	---	27.2	0.69	15.4	2.1	---	---
Haynes C-276	N.S.	0.25	3.6	1.0	---	32.5	---	8.2	16.2	0.4	---
Haynes 25	N.S.	0.07	---	15.4	---	7.1	---	14.9	---	0.3	---
ERCuA1-A2	N.S.	0.38	---	---	---	---	70.5	---	---	3.4	---
ERCu	N.S.	0.31	---	---	---	---	66.0	---	---	---	---

N.S. - Not specified
Pattee et al., 1978 (Ref. 70).

weight of deposited metal associated with the welding of carbon steel, stainless steel, high alloy, copper, and aluminum by the gas metal arc process. The results are shown in Table A13.

The fume generation rates for stainless steel and high alloy steel electrodes were relatively lower than the rates for carbon steel electrodes (Ref. 70).

The ER4043 aluminum electrode produced fumes at a lower rate than the ER5356 electrode; the ER5356 electrode contains magnesium, which oxidizes more easily and has a high vapor pressure. In contrast, the ER4043 electrode contains silicon, which has a much lower vapor pressure (Ref. 70).

The fume generation rates and the ratios of weight of fumes-to-weight of deposited metal for an ERCu and an ERCu A1-A2 electrode were comparable in magnitude to those associated with carbon steel electrodes (Ref. 70).

In the above study and others, the fume generation rate was found to depend upon a number of factors. The shielding gases used in gas metal arc welding have a pronounced effect on the rate of fumes produced during the welding process. A larger amount of fumes is produced with carbon dioxide than with an argon-5% oxygen mixture (Refs. 80 and 81), when both mild steel and low alloy steel are welded. The accelerated vaporization with carbon dioxide is due to its high thermal conductivity and high oxidizing power relative to argon. A minimal amount of electrode metal is vaporized in pure argon.

As observed during shielded metal arc welding, the fume generation rate during gas metal arc welding also increases gradually with increasing current (Refs. 70 and 81). Also, the rate of fume formation during gas metal arc welding increases with an increase in arc voltage, and the observed increase in the rate is much greater with argon than it is with carbon dioxide.

Gas Tungsten Arc Welding

In gas tungsten arc welding, fusion is obtained by heating with an arc between the base metal and a non-consumable tungsten electrode. Shielding is obtained from a gas or gas mixture; helium, argon, or a mixture of these two gases is frequently used. Depending upon the joint configuration, a filler metal may or may not be needed. In this process, the tungsten electrode serves only to maintain the arc. Filler metal, when used, is added in the form of a rod, and it does not form part of the welding circuit (Refs. 82 and 260). The electrodes used for gas tungsten arc welding are not usually pure tungsten. During electrode production, thoria (thorium oxide) in amounts of 1 to 2 percent is often added to tungsten to improve the arc initiation characteristics and arc stability, and to increase the current-carrying capacity of the electrode during welding. In some cases, 0.3 to 0.5 percent zirconium is used instead of thoria (Refs. 82 and 268).

The gas tungsten arc process is adapted to welding a

Table A13
Fume generation rates of various gas metal arc welding electrodes

Electrode type	Current range, A	Shielding gas	Fume generation rate, g/min	Weight of fumes-weight of deposited metal, g/kg
Carbon steel				
E70S-3	260-290	$Ar+2O_2$	0.41-0.46	4.97-5.68
	205-225	$Ar+9CO_2$	0.41-0.49	6.39-8.34
	320-330	CO_2	0.45-0.51	3.09-3.31
E70S-5	325-345	CO_2	0.40	2.61
Stainless and high alloy steel				
ER316	165-175	$Ar+O_2$	0.04	0.58
ERNiCu-7	250-260	Ar	0.16	2.02
Inconel 625	190-195	Ar	0.06	0.87
Haynes 25	200-205	Ar	0.08	1.38
Haynes C-276	165	Ar	0.39	6.98
Aluminum				
ER4043	160-165	Ar	0.11-0.27	5.6-15.74
ER5356	150-165	Ar	1.41-1.75	64.94-79.72
Copper				
ERCu	205-210	Ar	0.30	4.93
ERCu Al-A2	210-215	Ar	0.47	8.12

Pattee et al., 1978 (Ref. 70)

wide range of materials, such as carbon steel, low and high alloy steels, aluminum and aluminum alloys, magnesium and its alloys, copper, copper-nickel alloys, brasses, silver, and many others (Ref. 268). The fumes originate mainly from the base metal and external filler metal when the latter is used. Vorontsova (Ref. 91) compared the total fume concentrations emitted during the welding of aluminum and aluminum alloys by the shielded metal arc, gas metal arc, and gas tungsten arc processes. The shielding gas used was argon. The total fume concentrations in the breathing zones of the welders ranged from 1.3 to 4 mg/m^3 during argon tungsten arc welding; 12.6 to 54.6 mg/m^3 were found with the argon metal arc process. A still higher concentration of fumes was formed during shielded metal arc welding.

Pattee et al. (Ref. 70) studied the fume generation rates from welding stainless steel (200 to 215 A, dc, negative polarity, AWS EWTh-Z electrode) using a 2.38 mm 308L external stainless steel filler wire, and from aluminum (250 A, ac, AWS EWP electrode) using a 2.38 mm 5356 aluminum filler wire, with argon in both cases. The fume generation rates were 2.5 mg/min and 6.5 mg/min, respectively.

Heile and Hill (Ref. 81) used 3 mm and 5 mm thoriated tungsten electrodes to weld steel with argon as a shielding gas. No filler metal was used. The fume formation rate for all currents between 50 and 450 A was zero.

Another potential concern in gas tungsten arc welding is the use of thorium in tungsten electrodes. It is added to some electrodes at a concentration of 1 to 2 percent. Although the electrodes used for gas tungsten arc welding are considered to be nonconsumable, in practice, some loss of metal does occur. The rate of electrode consumption may vary from 0.1 to 60 mg/min depending on current, electrode diameter, and welding technique (Ref. 269). During this process, some thoria (thorium oxide) becomes airborne.

Breslin and Harris (Refs. 269 and 270) measured the radioactivity and thorium generated during the welding of aluminum, mild steel, and stainless steels with 3/32 or 5/32 in. electrodes containing 1 or 2 percent thorium. The rate of thorium production was found to be proportional to the loss of weight of the electrode. All samples collected 12 in. from the arc or beyond produced less than 20 disintegrations/min/m^3 of air. This is to be compared with the maximum allowable level, in force by the AEC at that time, of 70 disintegrations/min/m^3 for uranium.

Plasma Arc Welding

Plasma arc welding is an extension of the gas tungsten arc process in which fusion is obtained by heating with a plasma or constricted arc established between a non-consumable electrode and the base metal. The arc is formed at a relatively small orifice through which a stream of air, argon, helium, hydrogen, or a mixture of these gases flows (Refs. 82 and 271). Since plasma arc welding is very similar to the gas tungsten arc process, the fume hazards associated with the two processes are generally similar.

Submerged Arc Welding

This process is widely used to weld relatively thick plates at high metal deposition rates. It is used most commonly for welding plain carbon and low alloy steels. High alloy ferrous metals and certain nonferrous metals are also welded by this method (Ref. 82). Submerged arc welding differs from other arc welding processes in that the arc is not visible, but is submerged under the flux. The end of the electrode and the molten weld metal are surrounded and shielded by an envelope of molten flux, upon which a layer of unmelted flux is superimposed (Ref. 82). The fumes originate from the base metal, the electrode, and the flux. Since the arc and the weld metal are protected by flux, the amount of fume emitted is minimal when compared to the shielded metal arc, gas metal arc, or flux cored arc processes. The major hazard present in submerged arc welding is the formation of fluoride fumes from the flux (Ref. 82).

Byczkowski et al. (Ref. 272) investigated the fumes evolved during submerged arc welding operations in a Polish shipyard. The welding was performed at a current of 650 to 700 A and a voltage of 38 V. A Polish made flux was used containing 43.3% SiO_2, 37.9% MnO, 5.2% CaO, 4.5% Al_2O_3, 3.2% CaF_2, 0.65% Fe, 0.04% P, and 0.04% S. Welding was performed at a rate of about 20 meters per hour. The electrode was 5 mm in diameter, consumed at a rate of 10 kg/hour. The flux was used at a rate of about 18 kg/hour. The concentration of the fumes was determined in welder's working area and at distances of 6 m and 12 m from the working area. The results of fume concentrations are given in Table A14, and an analysis of selected constituents is given in Table A15.

Oxygas Welding

The fumes generated in this type of welding originate in the base metal, filler metal, and fluxes; however, the temperatures produced are much lower than during arc welding, and the fume levels compared to the latter are minimal (Ref. 82). Significant levels of fume may be generated during the welding of galvanized steel or an alloy containing a volatile metal (Ref. 82).

Table A14
Concentrations of fumes during submerged arc welding in a shipyard production area

Sampling location*	Fume concentration, mg/m^3
Welder's working area	12.4
6 m from welder's working area	9.1
12 m from welder's working area	7.2

*During analysis, shielded metal arc welding was also in operation in the shipyard

Byczkowski et al, 1964 (Ref. 272).

Table A15
Analysis of fumes collected during submerged arc welding in a Polish shipyard area

Sampling location	Fume concentration, mg/m^3			
	SiO_2	FeO	Mn	F
Welder's working area	4.59	3.30	0.18	0.32
6 m from welder's working area	2.37	3.10	0.19	0.30
12 m from welder's working area	1.73	2.5	0.12	0.16

Byczkowski et al., 1964 (Ref. 272).

As evident from the foregoing discussion, each arc welding process has a fume formation rate that depends upon several factors. A definite conclusion as to which process produces more fumes in comparison to others cannot be drawn unless all the variables of the processes are considered. However, fume generation rate increases with current, and certain processes do require a higher current than others. Flux cored arc welding, for example, commonly employs higher currents than does shielded metal arc welding.

On the other hand, the electrodes in the shielded metal arc process are coated; flux cored electrodes contain similar flux compositions in their core, which add to the fume; whereas, the gas metal arc and gas tungsten arc processes use no fluxing or similar agents.

Lyon et al. (Ref. 97) reported approximate fume generation rates for some arc welding processes, and the results of this study are given in Table A16. Under these conditions, fume generation rate was highest for the gas shielded flux cored and shielded metal arc processes, followed by the gas metal arc and, finally, the gas tungsten arc process.

Table A16
Approximate fume generation rates during welding of mild steel by various process

Process	Electrode	Shielding gas	Current, A	Approx. fume generation rate, g/min
Gas tungsten arc	EWTh	Argon	50-300	<0.2
Gas tungsten arc	EWTh	Helium	50-275	<0.2
Gas metal arc	E70S-4	Carbon dioxide	90-350	0.2-0.4
Gas metal arc	E705-4	95% Argon + 5% Oxygen	150-350	0.2-0.4
Flux cored arc	E70T-1	Carbon dioxide	175-350	0.9-1.3
Shielded metal arc	E6013	---	100-200	0.8-1.2
Shielded metal arc	E7018	---	100-200	0.5-0.7
Shielded metal arc	E7024	---	100-200	0.3-0.5

Lyon et al., 1976 (Ref. 97).

Heile and Hill (Ref. 81) carried out an extensive study to provide fume generation rate comparisons between different arc welding processes. Shielded metal arc, gas metal arc, gas tungsten arc, and flux cored metal arc processes were studied over a range of welding currents. A comparison of the rates of fume generation from the various arc welding processes as a function of current is summarized in Figure A1. The amount of fumes generated during the gas tungsten arc process was zero; the highest fume levels were found in the case of flux cored arc welding using self-shielded electrodes. Unfortunately, the rates of fume formation were presented in mg/g of weld metal, so the faster the rate of welding, the higher the workroom air concentration would be.

The Extent of Fume Exposure

The extent to which welders may be exposed depends primarily upon several factors, such as the concentration of fumes and the duration of exposure to these fumes, the location of welding, personal protection, and engineering controls employed. The effects of welding fumes on the health of welders will also depend upon the composition of the fumes and the particle size distribution of that fume.

Fume Constituents

Some constituents of the fumes may pose more potential hazards than others, depending upon their inherent toxicity. Some of these constituents of special concern include silica, fluorides, copper, chromium, nickel, and manganese.

Silica

The principal source of silica in the welding fumes is from the coating of shielded metal arc electrodes and from the flux composition of flux cored arc electrodes. The coatings or the flux contain a high amount of silicon (5 to 30 percent) as silica, ferrosilicate, kaolin, feldspar, mica, talc, or waterglass (Ref. 273). Some low alloy steels and aluminum alloys also contain alloying silicon (Ref. 260). Thus, silica may be produced in the fumes resulting from the gas metal arc welding of low alloy steels or aluminum. The mechanism of formation of silica in the welding fumes has been studied by Heile and Hill (Ref. 81), who suggested that it involves elemental vaporization of silicon (from alloys), followed by condensation, and, finally, oxidation.

Buckup (Ref. 71) claimed that welding fumes contained only extremely fine amorphous silica, based upon unpublished results not further described. Crystalline silica (quartz), which is associated with silicosis, was not identified in the welding fumes he analyzed. In a recent study, Pattee et al. (Ref. 70) used scanning electron microscopy and x-ray diffraction methods to detect crystalline silica in the fumes from shielded metal arc (E7024 and E410-16 electrodes) and flux cored arc (E70T-4 and E70T-5 electrodes) welding. Crystalline phases of silica were not detected in the fumes.

Fluorides

The major source of fluorides in the fumes is also from the covering on shielded metal arc electrodes or the flux and slag composition of flux cored arc electrodes or submerged arc welding. The low hydrogen covered electrode and self-shielded flux cored electrodes contain large amounts of fluorspar (calcium fluoride). Some carbon dioxide shielded flux cored electrodes contain 18 to 20 percent calcium fluoride in their flux and 24 to 43 percent calcium fluoride in their slag compositions. The self-shielded flux cored electrodes contain 22 to 63 percent calcium fluoride in their flux and 47 to 76 percent in their slag compositions (Tables 7

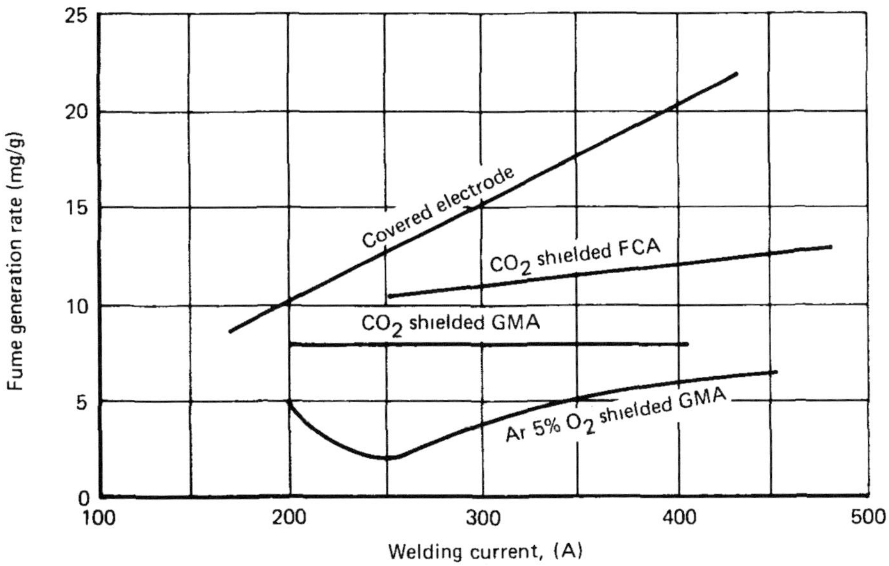

Heile and Hill, 1975 (Ref. 81).

Fig. AI-1 – Rate of fume generation for various arc welding processes as a function of current

and 8). The fluoride compounds present in the fumes from basic (low hydrogen) electrodes have been found to be in the range of 5 to 30 percent, calculated as fluorine. These are present as insoluble salts, mainly calcium fluoride, and soluble fluorine compounds, such as sodium or potassium fluorides. Up to 50 percent of the total fluorine content of freshly formed welding fumes are water soluble and are considered irritating to the mucous membranes. The presence of potassium in the coatings or flux affects the evolution of fluorides in the fumes. Thus, with a constant calcium fluoride content, but increasing potassium content, an increase in the evolution of fluorides is observed (Ref. 80). Pantucek (Ref. 273) reported that the fluoride ions in the welding arc or in the cooling cloud of fumes could react with hydrogen, potassium, sodium, copper, iron, silicates, manganese, titanates, and aluminates. Under the alkaline conditions of basic electrodes, only sodium fluoride, potassium fluoride, and calcium fluoride can exist as final products. Pantucek (Ref. 273) found that all fluorides were present as soluble salts in a study of fumes from basic electrodes. No fluorides in this study were found to exist as either silicon tetrafluoride or hydrogen fluoride. In a recent study of fumes from E70T-4 and E70T-5 flux cored electrodes, calcium fluoride was found to be present as a crystalline phase (Ref. 70). The E70T-1 flux cored electrode produced about 0.23 percent (by weight) of total fluorides in the fumes, of which 0.16 percent was water soluble (Ref. 70).

A covered electrode (E7018) was also investigated for fluoride content. It was observed that a large fraction of the fluorides in the fumes produced by this electrode was insoluble in water. Of a total 22 percent fluoride content, only 6 percent was water soluble (Ref. 70).

Copper

The highest copper concentrations are found when copper and its alloys are welded. Another minor source is from copper-coated gas metal arc electrodes.

Chromium

The welding of stainless steel and high alloy steels containing chromium presents a problem of chromium in the fumes, which has a low threshold limit value. Some chromium compounds have also been identified as carcinogens (Ref. 66).

Chromium can exist in various oxidation states from 0 to +6. Of these oxidation states, the best known are 0, +2, +3, and +6. The compounds in which chromium is in its +3 oxidation state (trivalent) are most stable, and the Cr^{+6} (hexavalent) compounds are important industrially (Ref. 274). From the standpoint of welding fumes, Cr^{+3} and Cr^{+6} may be more important. Some examples of Cr^{+3} compounds are chromic oxide (Cr_2O_3) and chromic chloride ($CrCl_3$). Examples of hexavalent chromium are chromium oxide (CrO_3), chromates, and dichromates (Ref. 274).

Stern (Refs. 78 and 83) compared the relative chromium content of welding fumes during the welding of stainless steel (15 to 25 percent Cr) and mild steel (unalloyed) by the shielded metal arc and gas metal arc processes. The total concentration of chromium in the welding fumes of mild steel was less than 0.05 percent in the shielded metal arc process and 0.005 percent in the gas metal arc process. Shielded metal arc welding of stainless steel with basic and rutile electrodes produced about 2.4 to 6.4 percent chromium, and the gas metal arc produced 9.8 to 13.8 percent chromium.

Of the chromium present in the fumes of stainless steel welding by the shielded metal arc process, 5 to 33 percent was found to be in its trivalent (Cr^{+3}) state or as metallic chromium, of which 100 percent was found to be insoluble in water. Sixty-seven to 95 percent of the total chromium content was found to be in its hexavalent state (Cr^{+6}), of which 0 to 13 percent was insoluble in water (Ref. 78).

In gas metal arc welding of stainless steel using argon or a mixture of argon and oxygen as a shielding gas, 98 to 99.86 percent of the total chromium present was found to be Cr^{+3} or metallic chromium, all of which was insoluble in water. The hexavalent chromium (Cr^{+6}) was present in 0.14 to 2.0 percent, of which 60 to 90 percent was insoluble in water (Ref. 78).

Stern (Ref. 78) suggested that the chromium present in the fumes from shielded metal arc welding was in the form of chromates and dichromates of sodium and potassium. The relative proportion of Cr^{+6} content of fume from gas metal arc welding depended upon the amount of oxygen in the shielding gas.

Virtamo and Tuomola (Ref. 84) determined the amount of hexavalent chromium in the fumes generated during shielded metal arc and gas metal arc welding of stainless steel (18 percent Cr). Using basic electrodes, the total chromium as Cr in the fumes was found to be in the range of 1.8 to 3.1 percent, depending upon the current, of which 0.57 to 2.2 percent was hexavalent chromium. The rutile electrode gave 3.6 percent total chromium, of which 2.5 percent was hexavalent. The proportion of hexavalent chromium was much lower than the level found by Stern, above. The gas metal arc process produced 0.40 to 0.96 percent chromium in the total fumes, of which 0.2 to 0.32 percent was hexavalent. All the electrodes used in this study contained chromium in the range of 18.5 to 20 percent.

Trivalent chromium (Cr^{+3}) has been considered to be of a low order of toxicity, while hexavalent chromium has been found to be more toxic (Refs. 275 and 276). Recent studies indicate that welding fumes containing hexavalent chromium have also shown mutagenic activity (Refs. 78 and 83). Some chromates have also been associated with an increased risk of lung cancer (Ref. 66).

Nickel

Nickel is present in stainless steel (typically 8 to 15 percent) and in nickel alloys. Some nickel compounds have been identified as carcinogens (Ref. 67), and a recent study (Ref. 88) shows that the welding fumes containing nickel, chromium, and other metals may be potentially mutagenic.

Manganese

Manganese oxide is used as a flux agent in the coatings of shielded metal arc electrodes, in the core of flux cored arc electrodes, and as an alloying element in the steel used in electrodes. In the study discussed previously (Ref. 70), manganese accounted for 0.3 to 8.8 percent and 1.0 to 13.5 percent of the total fume, respectively. Some special steels containing high manganese may produce high concentrations of manganese oxide in the fume (Ref. 80).

Other Fume Sources

Although the fumes produced during the welding process originate mainly from the electrodes, base metals, and the coatings or flux composition of the electrodes, in special cases some other materials may also constitute a considerable fraction of the welding fumes. For example, the welding of metals whose surfaces have been protected by galvanizing, plating, and painting can alter the fume generation characteristics of the base metals considerably; the fumes generated from these coatings may constitute as great or greater exposure potential than those generated by the welding process itself (Refs. 80 and 82). The most commonly encountered instance of this problem occurs with coatings containing zinc and lead (Ref. 80). The problems of lead and zinc are particularly serious in the shipbuilding industry where large quantities of zinc and lead based paints and coatings have been used to prevent corrosion. (Ref. 82).

Particle Size Distribution

The particle size distribution of the welding fumes is an important factor in determining the hazard potential of the fumes, since it is an indication of the depth to which particles may penetrate into the respiratory system and the number of particles that may be retained therein. Particles in the range of 1 to 7 μm in diameter represent the most serious hazard due to retention in the alveolar region (Ref. 85). Welding fume particles are essentially all less than 1.0 um in diameter (Refs. 70, 86, and 87).

Jarnuszkiewicz et al. (Ref. 87) measured the particle size distribution of the welding fumes in a shipyard hall and in a ship's double bottom where shielded metal arc welding using rutile electrodes was employed. The particle size distribution was determined by optical and electron microscopy under various climatic conditions (temperature, relative humidity, and air flow velocity). The results are given in Table A17.

Few particles were found in the range above 1 μm in the ship's double bottom only when the relative humidity was low (46 percent). At higher relative humidity (59 to 72 percent), the particles greater than 1.0 μm in diameter constituted 7 to 12 percent of the fume. The particle size distribution was not influenced by changes in temperature or air flow velocity.

Hedenstedt et al. (Ref. 88) carried out a study to determine mutagenic activity of the welding fumes generated during shielded metal arc and gas metal arc welding of stainless steel. During this study, it was observed that about 90 percent of the particles were

Table A17

Particle size distribution of the welding fumes in a shipyard hall and ship's double bottom

Working place	Ship's double bottom			Shipyard hall		
Temperature, °C	2.1	19.5	37.2	2.2	17.5	29.1
Relative humidity, %	69	72	46	66	62	59
Air flow, m/sec	0.07	0.04	0.06	0.35	0.30	0.86
Particle diameter, μm	Percent of particles					
Optical microscope						
<0.25	64.0	60.0	69.0	59.4	51.2	63.9
0.25-1.00	29.0	32.0	30.9	28.3	38.3	29.1
>1.00	7.0	8.0	---	12.3	10.5	7.0
Electron microscope						
<0.05	91.7	97.5	91.1	92.8	93.1	92.6
0.05-0.25	8.3	2.5	8.9	7.2	6.9	7.4

Jarnuszkiewicz, 1966 (Ref. 87).

smaller than 1 μm in diameter during welding of stainless steel. However, the particles of the fumes produced in the shielded metal arc process were relatively larger in size than particles formed in gas metal arc welding. Stern (Ref. 78) determined the particle size distribution in the fumes of the shielded metal arc welding of stainless steel and found the mass median diameter to be approximately 0.3 to 0.5 μm. The mass median diameter of the particles in the fumes of gas metal arc welding was found to be approximately 0.25 μm.

Heile and Hill (Ref. 81) determined the particle size distribution of fumes from gas metal arc and flux cored arc welding of steel. In both cases, most of the total fume particles had diameters of less than 1.0 μm. The gas metal arc welding was performed using a E70S-2 electrode (CO_2 shielded) and the flux cored arc welding was performed with a E70T-4 (self-shielded) electrode. The mean particle diameters were estimated to be 0.03 μm and 0.12 μm, respectively, from the gas metal arc and the flux cored arc welding fumes. The estimates of the diameters were based on extrapolations of least-square plots. These plots indicated that the log of the particle diameter was normally distributed. During gas metal arc welding, the mean particle size was compared for argon and carbon dioxide shielding gases. The mean particle size during welding with carbon dioxide was greater than with argon. This was explained on the basis that the rate of fume formation with carbon dioxide shielding was greater than with argon. It was concluded that in carbon dioxide atmospheres the particles had more time to grow in size than with argon.

Although it has been shown that most of the fume particles generated during arc welding are smaller than 1.0 μm in diameter, the particle can grow in size (agglomerate) with increased time. This agglomeration is largely due to thermal effects through particle collision. Thermal agglomeration is enhanced by the turbulent conditions resulting from heat generated in the welding process, thus increasing particle movement and chances for particle collision. During the shielded metal arc welding of low carbon steel, it was observed that the particles in the fume grow in size with time. The agglomeration of particles less than 0.93 μm was found to be greater than that of particles 5.5 μm or greater in diameter, in which very little growth occurred. Larger particles were most affected by gravity and settled out more rapidly than the particles of smaller size. It was concluded that approximately 2 minutes after welding ceases, the fume particles begin to grow in size when a significant number of particles less than 1.0 μm in diameter can agglomerate to form particles greater than 1.0 μm (Ref. 85).

Gases Produced During Welding

The major gases generated during welding processes are ozone, nitrogen oxides, carbon monoxide, and carbon dioxide. Other gases produced during welding may include phosgene, hydrogen chloride, and diacetyl chloride. They arise principally from the photochemical (welding light) oxidation of chlorinated hydrocarbon cleaning agents present in the atmosphere. Phosphine may also be produced from thermal decomposition of phosphate-containing metal coatings.

The gases produced during welding have several origins, depending upon the specific welding process (Ref. 70), and include: (1) shielding gases, (2) decomposition products of electrode coatings and cores, (3) reaction in the arc with atmospheric constituents, (4) reaction of ultraviolet light with atmospheric gases, and (5) decomposition of degreasing agents and organic coatings on the metal welded. These gases are discussed individually below.

Ozone

Ozone is an allotropic form of oxygen. It is produced during arc welding from atmospheric oxygen in a photochemical reaction induced by ultraviolet radiation emitted by the arc. There are two steps in this process (Ref. 277):

$$O_2 \xrightarrow[\text{radiation} < 210 \text{ nm}]{\text{ultraviolet}} 2\ O$$

$$O + O_2 \longrightarrow O_3$$

The reaction is induced by radiation of wavelengths shorter than 210 nm (Ref. 277). At wavelengths shorter than 175 nm, the absorption of ultraviolet light by oxygen is so effective that virtually none of this radiation penetrates further than a few centimeters beyond the arc; the effectiveness decreases with increasing wavelength (Ref. 89).

At wavelengths in the range of 220 to 290 nm, ozone absorbs ultraviolet energy and decomposes to reform diatomic oxygen (Ref. 28). It is also thermally unstable and decomposes to give diatomic oxygen; this process is slow at room temperature, but is accelerated greatly by heat (Ref. 28).

The rate of formation of ozone depends upon the wavelengths and the intensity of ultraviolet light generated in the arc (see the section on "Radiation" in Chapter 1), which in turn is affected by the material being welded, the type of electrode used, the shielding gas (if used), the welding process, and the welding variables, such as voltage, current, and arc length (Ref. 90).

Steel (Ref. 93) reported a survey of about forty shipyards employing shielded metal arc, submerged arc, and gas shielded arc welding processes, and ancillary processes such as flame cutting. The concentrations of ozone were found to range from 0.1 to 0.6 ppm (mean 0.35 ppm). The current OSHA standard for ozone is 0.1 ppm (see Appendix B for standards).

Variation of Ozone Levels with Process

Gas shielded arc welding (GMAW, GTAW) processes present a much greater problem from the point of view of ozone production than do shielded metal arc (Refs. 28, 91, and 92) or flux cored arc (Ref. 28) processes.

Vorontsova (Ref. 91) compared the amounts of ozone produced during the welding of aluminum and aluminum-magnesium alloys using the gas metal arc, gas tungsten arc, and shielded metal arc processes. No detectable amounts of ozone were found in the case of shielded metal arc welding. The comparison of gas metal arc and gas tungsten arc welding using argon as a shielding gas revealed that the quantity of ozone produced during gas metal arc welding was approximately four times greater than that produced with the gas tungsten arc process.

Lunau (Ref. 28) sampled for ozone at a distance of 6 in. from various welding arcs. Low levels (0.12 to 0.24 ppm) were detected during the welding of mild steel with either the shielded metal arc or flux cored arc process. Higher levels (0.27 and 2.1 ppm) were found when GTAW was used with mild steel. The highest levels (2.3 to 14.5 ppm) were attained when the argon shielded metal arc was employed to weld aluminum or aluminum alloys (Ref. 28). Although current affects UV irradiance, it was not a contributing factor to this difference since the gas shielded welding was performed at a lower current than either the SMAW or FCAW (Ref. 28). Although the greater production of ozone by GMAW (relative to GTAW) is corroborated by Frant (Ref. 89), Fay et al. (Ref. 278) found that under certain conditions the reverse obtained. One possible explanation for the higher production of ozone from gas shielded processes (relative to SMAW or FCAW) is the higher level of fume produced by the latter (see "Formation of Fumes" in this Appendix), which tends to block the emission of ultraviolet light; the effect of fume level on UV irradiance is discussed in detail in this Appendix under "Radiation from Arc Welding." Metal and shielding gas also contribute to arc spectral differences (Refs. 28 and 29), which also accounts for some of this effect.

There is little information on the formation of ozone during plasma arc welding. One study by Hickish and Challen (Ref. 217) revealed no detectable ozone levels with the plasma jet process. Levels in excess of 0.1 ppm may be produced during plasma arc cutting processes (Ref. 279).

The formation of ozone during submerged arc welding should be negligible, since there is no visible evidence of the arc. The area around the arc is surrounded and shielded by molten flux.

The oxygas process does not present a problem of ozone production, since the flame is not hot enough to emit light of sufficient energy to generate ozone.

The remainder of the discussion will deal only with gas shielded arc processes.

Variation of Ozone Production with Distance from Arc

Since the ozone is produced by the action of ultraviolet light upon atmospheric oxygen, it will be formed outside of the arc, but at a decreasing rate with distance. Since the ozone, unlike fumes and most other gases, is not formed directly in the arc, this fact has consequences for industrial hygiene measures to reduce ozone levels.

Frant measured the rate of production of ozone at various distances from the argon shielded tungsten and argon shielded metal arcs, using a cell with a quartz window attached to a supply of pure air (uncontaminated by nitrogen oxides) that flowed into a sampler. The variation with distance is presented in Table A18. These are not absolute figures for the decrease in production rate with distance since the quartz window was not 100 percent transparent to ozone generating wavelengths.

Lunau (Ref. 28) measured the average ozone levels at 9 and 18 in. from the argon-2% oxygen shielded metal arc. When mild steel and aluminum were welded, the ratio of the ozone levels at the two distances were 2.9 to 1 and 13.9 to 1, respectively. Since the ultraviolet radiation is more strongly absorbed at shorter wavelengths (see "Radiation from Arc Welding" in this Appendix), Lunau considered the difference in the rate of fall-off with distance to be due to differences in spectral characteristics of the two arcs.

Table A18
Variation of ozone production rate with distance for argon shielded tungsten and argon shielded metal arc processes[a]

Process	Distance from the arc, cm	Ozone production rate, μm/min
Argon-tungsten arc[b]	20	6.8[d]
Argon-tungsten arc[b]	50	0.6[d]
Argon-metal arc[c]	20	383.0[d]
Argon-metal arc[c]	40	70.6[d]
Argon-metal arc[c]	60	20.4[d]
Argon-metal arc[c]	80	10.7

a. Frant, 1962 (Ref. 89).

b. Nonconsumable tungsten electrode at 30 V and 150 A.

c. Consumable 106 mm aluminum electrode on aluminum at 32 V and 300 A.

d. Flow rate of 2.39 l/min of air through cell.

Other Factors Affecting Ozone Production

The shielding gas, weld metals, and current each have a pronounced effect on the ozone generation rate. This effect, however, appears to be secondary, and is due to changes in ultraviolet light intensity and spectral variation.

Shielding Gas and Weld Metal

Shielding gases and metals produce their effects by causing arc spectral changes. Those that increase the intensity of radiation at wavelengths shorter than 210 nm will increase the ozone generation rate (see "Radiation" in Chapter 1).

It appears that the highest levels of ozone are generated by the use of argon shielding with aluminum in the gas metal arc process (Refs. 28 and 89). Frant (Ref. 89) measured the ozone generation rate in GMA welding of aluminum (base and electrode) with argon, steel with argon, and steel with carbon dioxide; the ozone generation rates were 300, 33, and 7 μg/min, respectively, at a distance of 30 cm from the arc, using the same experimental setup described previously. These data are presented in Table A19. Frant attempted to measure the spectra produced by the argon metal arc with aluminum and by the carbon dioxide metal arc with steel. Whereas the former revealed intense emission lines at 184 and 191 nm, only low intensity broad band radiation appeared at wavelengths shorter than 210 nm in the latter.

Table A19
Effect of shielding gas and weld metal on ozone generation rate for the gas shielded metal arc process[a]

Shielding gas	Electrode	Base metal	Ozone production rate, μg/min
Argon	Aluminum	Aluminum	300
Argon	Steel	Steel[b]	33
CO_2	Steel	Steel[b]	7

a. Using 1.6 mm electrodes at 300 A and 37 V; sampling at 30 cm from the arc; Frant, 1963 (Ref. 89).

b. Presumably mild steel.

Frant (Ref. 89) also claimed from previous work that for welding aluminum and copper with helium as a shielding gas generated ozone at 15 to 20 times lower levels than when argon was used.

The presence of alloying elements can play an important role in the ultraviolet light spectrum and, consequently, in ozone generation. Lunau (Ref. 28) compared the ozone levels when welding pure aluminum, an aluminum alloy containing 5 percent magnesium, and an alloy containing 5 percent silicon. The results are summarized in Table A20. The addition of magnesium suppresses ozone production, while silicon augments it. A comparison of the spectra from the welding of aluminum and the welding of the magnesium alloy reveals that the addition of magnesium suppresses the aluminum line at 185 nm, possibly because of its higher vapor pressure and lower ionization potential in comparison to aluminum.

Current and Other Factors

Increasing current increases the ultraviolet irradiance (see "Radiation" in Chapter 1), which increase may be expected to raise the ozone generation rate. Lunau (Ref. 28) found that ozone levels 6 in. from the arc in both GTA and GMA welding did increase with increasing current, but not as rapidly as would be predicted from the increase of UV irradiance with current.

Table A20
The effect of alloying elements on ozone generation in the gas metal arc process 6 in. from the arc[a]

Alloy	Shielding gas	No. of tests	Ozone level
Pure aluminum	Argon-2% O_2	6	6.1
Al-5% Mg	Argon	6	3.1
Al-5% Mg	Argon-2% O_2	6	2.3
Al-5% Si	Argon	12	14.2
Al-5% Si	Argon-2% O_2	12	14.5

a. Sampling over 3-5 min. period, commencing 1 min. after initiation of welding. All runs were at 300 A.

Lunau, 1967 (Ref. 28).

Lunau (Ref. 28) also found that there was no effect of doubling the argon flow rate from 30 to 60 ft^3/hr on ozone concentration (5.1 and 4.5 ppm, respectively) 6 in. from argon shielded welding of aluminum (1/16 in. diameter electrode).

No appreciable difference was noted whether pure argon or argon with 2 percent oxygen was used as a shielding gas (Ref. 28). This is evident from examination of Table A20.

Nitrogen Oxides

The oxides of nitrogen are formed during welding processes by the direct oxidation of atmospheric nitrogen at high temperatures produced by the arc or flame (Refs. 93 through 95). The first reaction to take place is the formation of nitric oxide (NO) from nitrogen and oxygen (Ref. 95).

$$N_2 + O_2 \rightleftharpoons 2\,NO$$

The rate of formation of nitric oxide is not significant below a temperature of about 1200° C, but increases with increasing temperature. After dilution with air, nitric oxide can react further with oxygen to form nitrogen dioxide.

$$2NO + O_2 \rightleftharpoons 2\,NO_2$$

Above 500° C, the equilibrium of this reaction lies far to the left, but the equilibrium lies entirely on the side of nitrogen dioxide at room temperature. However, the rate of formation of nitrogen dioxide from nitric oxide is slow enough that both will be present in the atmosphere (Ref. 95). The stimulus for the production of nitric oxide from the atmospheric elements is heat, and temperatures in the oxygas flame (Ref. 95) and the welding arc (Ref. 280) are sufficient for this reaction to be significant.

In a survey of shipyards in 1968, Steel (Ref. 93) found nitrogen dioxide at an average concentration of 10.3 ppm. The processes employed in the shipyards were cutting and shielded metal arc welding. Data on individual processes are presented below.

Shielded Metal, Flux Cored, and Gas Metal Arc Welding

Alpaugh et al. (Ref. 92) in 1968, measured nitrogen dioxide produced by welding 3/4 in. plates of mild steel by the shielded metal arc, flux cored arc, and gas metal arc processes. Grab samples were taken during the approximately five minute welding periods, both inside and immediately in front of the helmet on a dummy (breathing simulated by air pumps). The samples were analyzed by the phenol-disulfonic acid method. Nitrogen dioxide levels ranged from 0.33 to 1.49 ppm inside the helmet, and 0.50 to 7.19 ppm outside, as can be observed in Table A21.

Gas Tungsten Arc Welding

The concentration of nitrogen dioxide was measured by the phenol-disulfonic acid method 1 to 2 ft from a gas tungsten arc struck on a water-cooled copper block. Sampling time was not stated. The results are summarized in Table A22. In all cases, a higher concentration of nitrogen dioxide was obtained when argon rather than helium was used as the shielding gas (Ref. 278).

Plasma Arc Welding

In 1968, Opris and Ionescu (as discussed in Ref. 82) reported concentrations of nitrogen dioxide ranging from 0.06 to 0.08 ppm, 1 to 4 meters from the arc during plasma arc welding; the maximum concentrations occurred 1.75 to 2.5 meters away. However, during plasma arc cutting, the concentration of nitrogen dioxide was 21 and 16 ppm, 1 meter and 7 meters from the arc, respectively.

Oxygas Welding

During the welding of mild steel in an unventilated 2000 cu ft room, Fay et al., 1957 (Ref. 28), took grab samples 2 ft away from and 2 ft above an oxyacetylene flame, in the breathing zone of the welder, and 6 in. from the flame. In the first two regions, the nitrogen dioxide concentration did not exceed one ppm; in the latter it was approximately 8 ppm. Morley and Silk (Ref. 110) measured nitrogen dioxide levels in shipbuilding and ship repair yards by "indicator tube" at the welder's breathing level while oxygas welding. Averages at 5 sites ranged from 9.0 to 16.5 ppm; they felt that a hazard exists when flames are used in confined or semi-confined spaces without adequate precautions.

Carbon Dioxide and Carbon Monoxide

Carbon dioxide and carbon monoxide are formed by the decomposition of organic compounds in electrode coatings and cores, and from inorganic carbonates in coatings. They are formed in the oxyacetylene flame, and carbon monoxide is formed by the decomposition of carbon dioxide used in the gas shielded metal arc

Table A21
Nitrogen dioxide concentrations produced inside and outside a helmet by various welding processes[a]

Process	Shielding gas	Electrode	No. of tests	Nitrogen dioxide (ppm) Inside helmet, avg (range)	Outside helmet, avg (range)
SMAW[b]	---	E7018	5	1.49 (0.43-2.55)	1.62 (1.07-2.34)
		5/32 in. x 14 in.	9	0.9 (ND[c]-1.7)	3.6 (0.43-7.68)
			10	0.8 (0.2-1.1)	1.3 (0.4-2.8)
			11	0.37 (ND[c]-0.85)	0.80 (ND-1.70)
FCAW[e]	CO_2	E70T-2	9	0.54 (ND[c]-0.91)	2.3 (1.7-4.25)
	---	E70T-4	5	1.36 (ND[c]-2.55)	7.19 (3.4-10.8)
GMAW[d]	Ar-2% CO_2	E60S-3	10	1.1 (ND[c]-3.0)	2.1 (0.2-4.5)
	Ar-25% CO_2	E60S-3	11	0.33 (ND[c]-1.07)	0.50 (0.21-0.85)

a. Alpaugh et al., 1968 (Ref. 92).

b. 28 V and 215 A.

c. ND = not detectable.

d. 32 V and 200-400 A.

e. Presumably same as b.

Table A22
Levels of nitrogen dioxide measured one to two meters from a gas tungsten arc[a]

Shielding gas	Nitrogen oxides, ppm
Helium	0.3[b]
Helium	0.5[b]
Helium	0.3
Argon	2.5
Argon	3.0[b]
Argon	3.0[b]
Argon	2.5

a. At 110 A, dcsp, 3/32 in. electrodes without filler metal and 15 ft^3/hr of shielding gas. Terry and Guther, 1953 (Ref. 278).

b. Flow rate of shielding gas doubled.

process. Standards for air concentrations are presented in Appendix B.

In shipyards employing shielded metal arc welding, concentrations of carbon monoxide were found to be below the threshold limit value of 50 ppm. However, higher concentrations were detected near the arc when carbon dioxide shielded metal arc welding was in operation (Ref. 93).

In 1960, Hummitzsch (Ref. 263) measured the concentrations of carbon monoxide produced during manual and automatic carbon dioxide shielded metal arc welding. In both cases, levels of carbon monoxide were higher than 100 ppm as far as 2 feet from the arc. Even in a ventilated room, the concentration reached 150 to 200 ppm after 10 minutes of welding. Likewise, Erman et al. (Ref. 283) measured the concentrations of carbon monoxide during the carbon dioxide shielded arc welding of steel in shipyard construction. The concentrations were determined at relatively open sections of the ship structure. It was found that the amounts of carbon monoxide produced were in the range of 0.17 to 0.23 g/min or 1.41 to 4.19 g/kg of the electrode consumed.

In submerged arc welding, the primary source of carbon monoxide is the carbon in the welding wire and steel being welded. The concentrations in a poorly ventilated area could read 100 ppm in about 15 to 30 minutes of welding (Ref. 82).

Gases Formed From Decomposition of Chlorinated Hydrocarbons

Chlorinated hydrocarbons are often used to degrease metal parts. When those metals contain residual degreaser and are welded, or the solvent is present in the atmosphere, the chlorinated hydrocarbons are decomposed by the ultraviolet light of the arc and, in some cases, by heat to produce phosgene, hydrogen chloride, dichloroacetyl chloride, and chlorine, depending upon the chemical nature of the hydrocarbon (Refs. 284 through 286). Trichloroethylene is the most commonly used solvent for degreasing. Others include tetrachloroethylene and methyl chloroform.

Dahlberg (Ref. 287) studied the decomposition of trichloroethylene when a degreaser was placed 15 to 20 m from SMA, GMA, and GTA welding operations. Dichloroacetyl chloride and phosgene, formed by photochemical oxidation, were the main products, and levels measured by gas chromatography never exceeded 10.4 and 3 ppm, respectively. Sampling was usually 30 cm from the arc. Dahlberg (Ref. 287) stated that the rate of formation of dichloroacetyl chloride was approximately five times greater than that of phosgene. The data suggest that the concentration of phosgene and dichloroacetyl chloride were higher during gas metal arc welding than during gas tungsten arc welding.

The following reactions led to the observed decomposition products:

$$CHCl\,CCl_2 + 1/2O_2 \xrightarrow{\text{light}} CHCl_2CClO$$

$$CHCl\,CCl_2 + O_2 \xrightarrow{\text{light}} COCl_2 + HCl + CO$$

Therefore, amounts of hydrogen chloride and carbon monoxide, equivalent to the phosgene, should also be formed (Ref. 284).

The photochemical oxidation of perchloroethylene produces equal quantities of trichloroacetyl chloride and phosgene, according to Dahlberg (Ref. 284). Phosgene is formed from perchloroethylene at a faster rate than from trichloroethylene, because of the higher quantum yield of the former oxidation (Refs. 286 and 288).

Likewise, methyl chloroform produces hydrogen chloride and phosgene in a ratio of about 5 to 1, respectively (Refs. 284 and 285). Carbon tetrachloride and trichloroethylene produced comparable amounts of phosgene.

Radiation From Arc Welding

Electromagnetic radiation in the ultraviolet, visible, and infrared portions of the spectrum is emitted by most arc welding processes. The eye can be adversely affected by radiation in all three energy ranges; the skin is also susceptible to the effects of UV light. Exposure to arc radiation is all the more important, because, besides the welder himself, the welder's helper and others near the welding site are also at risk.

Both the intensity of light and the qualitative differences in the spectra produced in various arc welding processes will affect the welder's exposure. Also, ozone is produced by light of only certain wavelengths. These and other factors affecting exposure are presented in the following paragraphs.

Spectral Variation

The welding arc generates line spectra characteristic of the materials involved in the process superimposed upon a continuum of radiation (Ref. 39). Radiation arises from the arc and the molten metal pool (Ref. 289). Based upon the assumption that the arc and pool were black body radiators, Van Someren and Rollason (Ref. 290) calculated that the distribution of energy from shielded metal arc welding of an iron-based material using 4 gage covered electrodes at 280 A would be as follows:

UV	200-400 nm	5%
Visible	400-750 nm	26%
IR	750-1300 nm	31%
	beyond 1300 nm	28%

These figures have been often quoted (Refs. 46, 90, and 150); however, these data are inapplicable to other conditions and processes to the extent that radiation shorter than 200 nm is produced, the arc is not a black body radiator, and the temperature of the arc and pool are respectively greater (Ref. 150) than the 4650° K and 1930° K assumed. For example, Glickstein (Ref. 291) has shown that the temperature of a 100 A argon tungsten arc with a 2 mm long arc and uncooled workpiece was greater than 8000° K.

The spectra from a number of arc welding processes have been obtained. The AWS Committee on Safety and Health, in cooperation with the U.S. Army Environmental Hygiene Agency and Union Carbide Corp., sponsored a study to measure the spectra produced by various arc welding and cutting processes. They studied shielded metal arc, gas metal arc, gas tungsten arc, flux cored arc, and plasma arc welding processes, as well as plasma arc cutting, on mild steel and aluminum. Spectra covering 200 to 800 nm have been published (Refs. 97 and 292). Dahlberg (Ref. 293) also has obtained spectra for shielded metal arc and gas metal arc welding of carbon steel, stainless steel, and aluminum.

The spectral lines arise principally from ionization of filler and base metals and any shielding gas (Refs. 293 and 294). Addition of any volatile alloying element with a low ionization potential will alter the spectrum. Lunau (Ref. 28) showed that when an alloy containing 5 percent magnesium, instead of pure aluminum, was welded with the argon shielded arc process, the strong lines around 185 nm were suppressed, and this was considered due possibly to the lower ionization potential of the vapor.

Also, Frant (Ref. 89) has shown that, qualitatively, the spectrum from the welding of aluminum with argon is quite different from the CO_2 shielded welding of steel (not further described); in the former, there were intense emissions around 184 and 191 nm, whereas, there was only a low level background in the region from 170 to 200 nm in the latter, and this is the region important for producing ozone.

Virtually no light of wavelength shorter than around 175 nm should strike the welder. Oxygen in the air absorbs UV light of wavelengths shorter than 200 nm. At wavelengths shorter than 175 nm, this process is so effective that the intensity of this radiation produced should be reduced essentially to zero after passing through a few cm of air (Refs. 28 and 89).

Obstruction of Radiation

The intensity of at least the radiation in the ultraviolet portion of the spectrum is attenuated with distance at a rate faster than predicted by the inverse square law (Ref. 295). This is due principally to obstruction by fumes. The movement of fumes may also account in part for the rapid changes in intensity encountered when measuring arc emission spectra (Ref. 41). In one experiment (Ref. 96), the average ultraviolet irradiance at 4.8 meters from a shielded metal arc welding operation, using a 6011 electrode at 380 A, was 1.5×10^{-6} W/cm^2 without fume removal, 77×10^{-6} W/cm^2 with moderate air flow fume removal, and 100×10^{-6} W/cm^2 with high air flow fume removal. At a distance of 19.2 meters from the arc, the values were 0.38×10^{-6}, 0.57×10^{-6}, and 4.5×10^{-6} W/cm^2. Overall, the irradiances were 12 to 100 times greater when a high rate blower (rate not specified), rather than natural ventilation, was used (Ref. 96). The slope of the line obtained for the decrement in UV radiant exposure with distance showed only a slightly higher attenuation with distance than would be expected by the inverse square law when the highest ventilation rate was used. This residual attenuation might be due to the absorption of radiation shorter than 200 nm by oxygen to form ozone and the subsequent decomposition of this ozone by 220 to 290 nm light (see also "Radiation" in Chapter 1 for details).

Effect of Current

It is generally agreed that the UV irradiance increases with increasing current (Refs. 96, 97, and 293). Dahlberg (Ref. 293) observed this qualitatively in argon shielded welding (98% Ar, 2% O_2) of stainless steel. Lyon et al. (Ref. 97), however, showed that the increase in actinic ultraviolet (light between 200 and 315 nm) irradiance is roughly proportional to the square of the current for gas tungsten arc, gas metal arc (CO_2), and flux cored arc (CO_2) welding of mild steel. Visible luminance appears to increase at a slower rate with increasing current (Refs. 292 and 296).

Reflection

Exposure can also be enhanced by reflection of emitted radiation from other surfaces. Polished metal surfaces reflect over 85 percent of incident ultraviolet radiation, whereas surfaces coated with ZnO or TiO_2 paint should reflect less than 10 percent of incident UV radiation (Refs. 90 and 147).

Process Differences

Gas metal arc welding appears to produce the most intense ultraviolet radiation (Ref. 90). Dahlberg (Ref. 293) obtained spectra from the arc welding of carbon steel, stainless steel, and aluminum. Over the wavelength range from 190 to 290 nm, the gas metal arc process produced more intense radiation than did shielded metal arc welding, although the current used was slightly higher in the former process (180 or 200 as opposed to 150 A); for the latter, rutile, alkaline, and acidic covered electrodes were examined. The argon shielded arc (85% Ar + 2% O_2, 180 A on carbon steel) produced much more intense radiation over this range than did a CO_2 shielded arc (stainless steel, 200 A). Dahlberg considered this to be due to the more efficient heat transport by CO_2, which makes the argon shielded arc hotter.

Over the range of 190 to 290 nm, Dahlberg (Ref. 293) also showed that for shielded metal arc welding of stainless steel at 150 A, more intense radiation was produced when rutile rather than acidic covered electrodes were used, and this was more intense than when alkaline electrodes were used; Dahlberg states that the fume level produced by electrodes coated with these three types of material was in the reverse order.

At comparable current levels, less UV and visible radiation should be produced by flux cored arc welding than by the gas shielded process (Ref. 90) because of the higher obscuring fume levels produced in the former (see the above discussion and the discussion of fume formation). Under proper conditions, little or no UV or visible radiation will be produced by submerged arc welding unless the flux is broken momentarily; however, some infrared radiation will be emitted (Ref. 90).

Noise

Welding and cutting processes vary in the noise levels they generate. Using a Type 1 meter, Rodman et al. (Ref. 98) measured the noise intensity of several welding processes in an acoustically dampered chamber at 5 locations one meter from the arc (directly above, and 0.35 m above the arc and either in front of, behind, or on either side of the welder). The results are presented in Table A23. The most quiet was the gas tungsten arc process. The flux cored arc and shielded metal arc process noise levels varied considerably, depending upon the welding conditions. The air carbon arc process generated the most noise. In none of these measurements was the power source a contributor, since it was placed outside the chamber.

The plasma arc process emits broad band noise with no prominent frequencies and with intensities ranging from 80 to 91 dB (over the frequency range of 300 to 10 000 cps) at the position normally occupied by the operator, according to Hickish and Challen (Ref. 217).

Rudenko and Kudrya (Ref. 215) claimed that noise from the oxygas torch is intense in the 1000 to 8000 Hz region and A-weighted (according to sensitivity of the ear) noise exceeds 70 dB.

These measurements described above represent only noise generated by an individual process. Often, welders are also exposed to the noise generated by other welding units and by other equipment operations in the shop or plant, such as ventilators and electrical generators. (For further information, see the AWS publication, "Arc Welding and Cutting Noise.")

Table A23
A-weighted[a] sound pressure level ranges created by various welding and cutting processes while the arc was struck

	Filler/electrode		Base metal			
Process	Diameter, in. (mm)	Type	Thickness, in. (mm)	Steel type	Current, A	Range, dB
GTAW	3/32 (2.4)	308-L	1/2 (12.7)	stainless	130, 160, 190[b]	50-60
GMAW	0.045 (1.1)	E70S-3	3/4 (19.2)	mild	c	70-82
FCAW	3/32 (2.4)	308-L	1/2 (12.7)	stainless	130, 160, 190[b]	50-62
FCAW	3/32 (2.4)	E70T-1	3/4 (19.1)	mild	290, 460, 540[d]	70-86
FCAW	3/32 (2.4)	E70T-4	3/4 (19.1)	mild	260, 390, 515[e]	70-86
SMAW	5/32 (4.0)	6010 IP	3/4 (19.1)	mild	130, 160, 190	62-68
SMAW	5/32 (4.0)	6010 IP	3/4 (19.1)	mild	115, 150, 180	68-82
SMAW	5/32 (4.0)	E7018	3/8 (9.5)	mild	130, 160, 190	66-78
ACA	3/16 (4.8)	f	3/4 (19.1)	mild	g	96-116

a. Weighted according to the sensitivity of the human ear.
b. Argon at 30 ft^3/hr.
c. 210 A and 91% Ar, 9% CO_2; or 275 A and 98% Ar, 2% O_2, both at 50 ft^3/hr.
d. CO_2 at 50 ft^3/hr.
e. No shielding gas.
f. Copper covered.
g. 16, 24 and 32 lbs pressure.

Rodman et al., 1978 (Ref. 98).

Other Factors

Heat is also generated in welding operations. It would be expected to be more intense when welding or cutting in confined spaces (Ref. 291) or in working with metal that has been preheated to improve welding characteristics. Fabrics differ in their heat transmission properties; therefore, a welder's protective clothing may affect the intensity of exposure (Ref. 90). In addition to any direct effects, heat and exertional stress are important because they can increase an individual's susceptibility to the harmful effects of other agents (for example, increased respiratory activity due to heat [Ref. 297] and exertion may lead to inhalation of larger quantities of fumes and gases).

Appendix B

Table B1
OSHA Standard or threshold limit value for selected airborne contaminants and physical agents

Agent	OSHA Standard, mg/m³ (ppm), 8 hr time weighted average	Threshold[a] limit value
Fume components[b]		
Aluminum	NA[c]	NA[c]
Cadmium		
fume	0.1	
dust	0.2	
Calcium oxide	5	
Chromium		
soluble salts, as Cr	0.5	
metal, insoluble salts	1	
Copper fume	0.1	
dusts and mists	1	
Fluoride (as F)	2.5	
Iron oxide fume	10	
Lead and its inorganic compounds	0.2	
Magnesium oxide fume	15	
Manganese[d]	5	
Nickel (metal and soluble compounds)	1.0	
Silica (quartz)		
respirable	e	
total dust	f	
Nuisance dust		
respirable	5	
total	15	
Titanium dioxide	15	
Vanadium[d]		
V_2O_5 fume[d]	0.1	
V_2O_5 dust[d]	0.5	
Zinc oxide fume	5	

Table B1 (continued)
OSHA Standard or threshold limit value for selected airborne contaminants and physical agents

Agent	OSHA Standard, mg/m³ (ppm), 8 hr time weighted average	Threshold[a] limit value
Gases[b]		
Carbon dioxide	9,000 (5,000)	
Carbon monoxide	55 (50)	
Nitric oxide	30 (25)	
Nitrogen dioxide	9 (5)	
Ozone	0.2 (0.1)	
Physical Agents		
Noise[g]	90 dBA	
Ultraviolet radiation		
(200-315 nm)[h]		0.1 mW/cm²
(320-400 nm)[i]		1 mW/cm²

a. Provided when OSHA Standard has not been promulgated. Source: American Conference of Government Industrial Hygienists. TLVs: Threshold Limit Values for Chemical Substances and Physical Agents in the Workroom Environment with Intended Changes for 1977, American Conference of Government Industrial Hygienists, Cincinnati, OH, 1977, 94 p. (Ref. 231).

b. Source: 29 Code of Federal Regulations 1910.1000 (1976), (Ref. 307).

c. NA = not available.

d. The value is a ceiling value and not 8 hr TWA.

e. Exposure limit in mg/m³ = 10/(% SiO_2 (quartz) +2)

f. Exposure limit in mg/m³ = 30/(% SiO_2 (quartz) +2)

g. Source: 29 Code of Federal Regulations 1910.95 (1976) (Ref. 308).

h. Weighted according to the relative spectral effectiveness of the incident light. See reference in footnote a for details.

i. For periods greater than 1000 seconds; for a period less than 1000 seconds, total exposure shall not exceed 1 J/cm²

Appendix C
Chronic Lung Disease Assessment

Pathogenesis of Lung Diseases

Pulmonary fibrosis (Ref. 75) is a progressive condition in which the lungs become less distensible, or restricted, increasing the work of breathing. The patient with this condition may appear cyanotic and show severe dyspnea during activity. This is because oxygen exchange is slowed in the lungs due to thickened membranes separating the air sacs from the blood capillaries in the lungs, termed alveolar-capillary block. In severe cases, the strain on the heart pumping blood through the fibrotic lung tissue may lead to right heart failure (cor pulmonale).

Some substances associated with pulmonary fibrosis include crystalline silica (Ref. 69), beryllium, asbestos, talc and diatomite (Ref. 41), smoking (Ref. 298), and possibly ozone (Ref. 29). However, iron oxide deposits in the lungs have only rarely been claimed to evoke fibrosis (Ref. 41).

Pulmonary emphysema (Ref. 75) is a progressive chronic disease in which there may be a loss of lung tissue as well as airway obstruction. Inhaled dust particles most often cause "focal" emphysema, which is a selective distribution of abnormally dilated air spaces around small bronchioles. The work of breathing is increased; expiration is more difficult than inspiration, due to collapse of airways during expiration, leading to trapping of air in the lungs. There is no cough in most cases. Dyspnea develops insidiously. The loss of lung tissue leads to loss of the capacity to oxygenate the blood. Increased strain on the heart to pump blood through the emphysematous lungs can cause right heart failure.

Causes of chronic pulmonary emphysema include cigarette smoking (Ref. 298), and long-term inhalation exposures to ozone (Ref. 176), possibly nitrogen oxides (Ref. 41), cadmium fumes (Refs. 41 and 103), phosgene (Refs. 41 and 102), and possibly other agents.

Another type of pulmonary emphysema is a genetic familial disease caused by a deficiency of the serum enzyme α-antitrypsin.

Bronchitis (Ref. 75) is defined as airway inflammation due to inhaled substances. The chronic form results from prolonged inhalation of an irritant or combination of irritants. The diagnosis is based upon the presence of cough with or without sputum for at least three consecutive months a year for at least 2 consecutive years. Smokers' morning cough is one example. Dyspnea and wheezing also occur. A mucopurulent form of chronic bronchitis is diagnosed when sputum contains pus. As the disease progresses, the airways become more vulnerable to infections. Scarring can occur in small airways. A condition known as bronchiolitis fibrosa obliterans consists of severe scarring (fibrosis) that obliterates small airways. The patient with advanced chronic bronchitis may be unable to obtain sufficient oxygen, resulting in cyanosis and dyspnea. Increased strain on the heart to pump blood through the scarred lungs can result in cor pulmonale.

Some probable causes of chronic bronchitis include cigarette smoking (Ref. 75), nitrogen oxides, and sulfur dioxide. The disease appears to be prominent in industrialized countries, in areas having cool humid climates and heavy air pollution (Refs. 11 and 75), and in older populations (Ref. 75).

Chronic bronchitis and emphysema often develop together, but whether one causes the other is still debated. Chronic bronchitis is also associated with fibrosis of the lungs (Ref. 75).

Lung Function Testing

Pulmonary function (breathing) tests (Ref. 75) are designed to measure the volume of air the lungs can inhale or exhale forcefully or during relaxed breathing. Lung volume can be abnormally low when a disease process or tumor restricts complete lung expansion during inhalation or when lung elasticity is impaired by fibrosis. The rate of airflow can be abnormally slow due to a narrowing or obstruction of airways that occurs during an asthma attack, in chronic bronchitis, or in emphysema.

Lung function tests are insensitive measures of lung dysfunction. Test results may appear normal in the presence of other manifestations of disease.

The specific lung function indices presented in the literature reviewed in this report are described below.

Vital Capacity (VC): the volume of air in the lungs that can be completely exhaled following a deep inspiration. Any disease that reduces the ability of the lungs to expand will reduce VC. Pulmonary fibrosis, chronic bronchitis, and pulmonary emphysema, regardless of the causative agent(s), cause VC to decrease. Tall, thin individuals have higher VC than obese persons who are otherwise healthy. Athletes usually have elevated VC.

Residual Volume (RV): the volume of air in the lungs that cannot be exhaled, even with force. In diseases where air becomes trapped in the lungs, such as pulmonary emphysema, the RV increases. In diseases in which fibrosis prevents normal lung expansion, RV decreases.

Tidal Volume (TV): the amount of air inhaled and exhaled during relaxed breathing. Only very severe lung diseases alter this volume.

Total Lung Capacity (TLC): the total volume of air in the lungs after maximal inhalation; RV + VC. In diseases in which air becomes trapped in the lungs, TLC is increased; for example, emphysema. In diseases where fibrosis prevents normal expansion of the lungs, TLC is decreased.

One-second Forced Expiratory Volume ($FEV_{1.0}$): the maximum amount of air that an individual can forcefully exhale in one second after forcefully inhaling as much air as possible. About 92 percent of the air in the lungs, except RV, should be exhaled forcefully in 3 seconds. About 70 to 90 percent of this air normally can be forcefully exhaled in the first second. Diseases that trap air in the lungs, such as emphysema, reduce $FEV_{1.0}$ because of the narrowing of airways (increased airway resistance during expiration). Normal values are based on age, sex, and height. Less than 70 percent of expected normal values is abnormal (Ref. 10).

Maximal Mid-expiratory Flow Volume (MMFV) and Maximal Mid-expiratory Flow Rate (MEFR): after a forceful inhalation of as much air as possible, the patient forcefully exhales, as for $FEV_{1.0}$. During the middle 50 percent of exhalation, the air flow rate and volume of air are measured. As for $FEV_{1.0}$, MMFV and MEFR are decreased by diseases that increase airway resistance of the breathing passages.

Peters et al., 1973 (Ref. 8), noted that pulmonary irritants (such as ozone and NO_2) could theoretically cause abnormal increases in lung air volumes (characteristic of pulmonary emphysema), while fibrogenic dusts (such as crystalline silica, asbestos, and, possibly, aluminum fumes) could cause decreases in lung air volumes due to fibrosis. The interpretation of lung function studies in welders can therefore be extremely complex.

Appendix D
Details of Experimental Animal Studies

This part of the report reviews the available literature relating to animal exposures to welding fumes, gases, or both from many different types of electrodes and welding processes. Most of the studies were done in Russia and Poland. Many are over 30 years old. The quality of reporting for some studies is especially poor; strains, ages, and numbers of experimental animals are often omitted, as is duration of exposure and fume content when an unfamiliar type of electrode is being used.

The Appendix is divided into sections that discuss acute and chronic effects produced by various routes of administration. While inhalation studies predominate, some experiments on intratracheal, intraperitoneal, and subcutaneous administration of suspensions or solutions of welding fumes are included. The effects of exposure to arc welding radiation are reviewed. Separate sections cover the problems of carcinogenicity, metal fume fever, and lung infections in relation to welding exposure (influenza and tuberculosis).

Because of the lack of complete reporting of experimental procedures and results in most of the papers reviewed in this Appendix, the value of any conclusions drawn by the various authors must be treated accordingly. Much work needs to be done to determine the potential animal toxicity of welding fumes and gases from modern electrodes and processes used in the United States and Europe.

Inhalation of Welding Fumes and Gases

This section contains the largest volume of available literature concerning animal experiments. It is broken down by species: rats, rabbits, guinea pigs, mice, and cats were included. No experiments in dogs or monkeys were reported. The poor experimental design leads to the conclusion that the effects of inhalation exposures should be more fully investigated in the future.

Tables 36 and 37 summarize the results of acute and chronic inhalation exposures in experimental animals. They appear in Chapter 3 of the main text.

Effects in Rats

In this section, inhalation experiments on rats are reviewed. Effects of welding fume and gas inhalation included mortality (Ref. 234), gastrointestinal tract inflammation (Ref. 242), pulmonary edema and hemorrhage (Ref. 235), pulmonary fibrosis (Refs. 19 through 21), pulmonary irritation (Refs. 235 through 238 and 300), reduced rate of weight gain (Refs. 19, 192, 234, and 236), methemoglobinemia (Ref. 236), and central nervous system abnormalities (Ref. 19). The effects of copper, copper and nickel, aluminum, and aluminum-magnesium welding fumes are also reviewed. Pulmonary fibrosis was noted in every study (Refs. 22, 23, 239, and 240). No incidental findings of lung tumors were reported in this literature.

Other effects noted in rats concern liver function (Ref. 245) and reproduction (Refs. 243 and 244).

General Toxicity and Lung Effects

This discussion is divided into acute and chronic exposures in rats.

Acute and Subacute

The acute toxicity of three types of electrodes was determined by Migai and Norkin, 1965 (Ref. 234), by exposing 10 to 12 rats to high concentrations of welding fumes produced by burning 1.2 to 2.4 kg of nickel-chromium containing electrodes (used in stainless steel welding) and recording the number of animals that survived. Fume concentrations were not reported. (Table

D1 lists fume content.) Electrode UONI-13/45 fumes were least toxic among the three.

Table D1
Fume content from electrodes 606/11 and UONI-13/45

Component	606/11	UONI-13/45
	(mg/m^3)	(mg/m^3)
Dust	280-330	250-320
MnO_2	20.2	6.8
CrO_3	22.5	0.053
F	32.1	22.0
SiF_4	4	6
	(ppm)	(ppm)
NO	17	24
NO_2	1	3
CO	15	30
HF	22	32

Welding fumes generated by burning twice as much of UONI-13/45 as the other two electrodes produced no lethal effects for up to 20 days. Results are shown below.

Type of electrode	No. rats	Electrode burned (kg)	Number of rats dead after 1 to 20 days of exposure: 1	2	3	4	5	6	7	8	9-20
606/11	11	1.2	0	2	3	3	1	1	1	-	-
981/15	12	1.2	0	0	0	2	5	1	2	2	-
UONI-13/45	10	2.4	0	0	0	0	0	0	0	0	0

Hewitt and Hicks, 1973 (Ref. 237), and Hewitt and Hicks, 1972 (Ref. 238), exposed male albino CSE rats weighing 200 to 250 g to welding fumes containing 500 mg/m^3 iron, 380 mg/m^3 SiO_2, 95 mg/m^3 manganese, 7 mg/m^3 lead, 1.0 mg/m^3 each copper and antimony, 0.4 mg/m^3 cobalt, and 0.1 mg/m^3 chromium. Total fume was reported to be 1500 mg/m^3, with mean particle diameter of 0.15 μ. These welding fumes were produced by burning one rutile covered Phillips C18 electrode every 3 minutes. Two rats exposed to fumes for 30 minutes and 7 exposed for 4 hours were autopsied 24 hours later. Brown discoloration of lungs and yellow stomach discoloration were noted, but there was no hemorrhaging, inflammation, or vascular congestion in lungs or gastrointestinal tract. The heart and major blood vessels, liver, kidney, other viscera, and stomach mucosa appeared normal. Organ weights were considered to be normal. Blood and urine analyses showed no abnormalities.

Histologically, the lungs contained large numbers of granular pigmented macrophages in alveoli, alveolar ducts, and lower bronchioles. Slight alveolar epithelial thickening and peribronchial edema were noted. There was no abnormal fibrosis or interstitial leukocytic infiltrate.

Eight rats were exposed to 1500 mg/m^3 of welding fumes for 4 hours, then two at a time were killed 1, 7, 28, and 75 days later. Particulates were noted in macrophages in the lung parenchyma, but not in bronchioles. No abnormalities were detected in general appearance, weight, or histological examinations of organs. No growth impairment was evident.

In experiments by Von Haam and Groom, 1941 (Ref. 235), 3 rats (sex, strain, and weight not provided) were exposed for 6 hours to 1600 to 2600 mg/m^3 of welding fumes from A5 electrodes (17.6% Fe_2O_3, 9.5% Cr_2O_3, 10.6% Mn_3O_4, 16.8% CaO, 16.1% Na_2O, 5% Al_2O_3, 14.6% F). The rats showed pulmonary edema and respiratory tract irritation due to exposure. None died. Two rats exposed for 6 hours to welding fumes died of bronchopneumonia. Repeated exposures to welding fumes in groups of 3 rats, consisting of 1/2 hour of exposure in the morning and 1/2 hour in the afternoon for 2 months, led to edema, hemorrhage, and peribronchial nodules.

Harrold et al., 1940 (Ref. 192), exposed 105 albino rats (sex and age not given) to welding fumes and gases for 6 hours daily, 5 days a week, for up to 228 hours (34 days). The fumes contained up to 398 mg/m^3 iron oxide and up to 3.3 mg/m^3 manganese. Gases included up to 70 ppm nitrogen oxides, up to 32 ppm of ozone, and normal proportions of oxygen and carbon dioxide compared with room air. Over 90 percent of the rats survived the exposures, although body weight gains were slightly less in those exposed than in the control animals. Autopsies performed after various exposure durations showed pulmonary deposits of iron particles that migrated toward the lung periphery and tracheobronchial lymph nodes. No pulmonary edema was detected, and radiological examinations of the animals revealed no abnormalities.

McCord et al., 1941 (Ref. 236), investigated welding fume toxicity in rats, finding methemoglobin formation and weight loss to be the important toxic effects. Albino rats of the same age (± 10 days) were exposed in groups of 24 to fumes for 6 hours a day, 5 days a week, for 45 days. Fumes were produced by welding during 1/2 hour intervals. Fumes from the covered electrodes contained 8.4% SiO_2, 5.4% TiO_2, 5.0% MnO_2,and 79.0% Fe_2O_3. Evolved gases included over 20% O_2, 0.22% CO_2, 0.16 to 2.07 ppm ozone, and 24 ppm NO_2. Fume levels averaged 444 mg/m^3 Fe_2O_3, 15.5 mg/m^3 Mn, and 61 mg/m^3 SiO_2; traces of titanium and calcium oxides were detected.

Three of 24 exposed rats died. One that died after one day of exposure showed no gross pathology at autopsy. One that died 22 days into the exposure showed multiple minute lung abscesses and marked siderosis. The third, which died after 26 days of exposure, was not autopsied. Weight gains averaged 32 gm for controls and 2.9 gm for exposed rats. Radiologic examination showed no exposure related abnormalities. Complete blood count and hemoglobin levels were determined every 7 to 14 days during the exposure. No differences between exposed and control rats were noted, except that substantial quantities of methemoglobin were found in exposed rats. After termination of the exposure, methemoglobin levels returned to normal, as shown in Table D2.

Table D2
Methemoglobin levels in albino rats exposed to welding fumes and gases

	Males			Females		
Exposure (days)	4	11	4.3	4	11	43
Average methemoglobin levels (%)	2.6	3.0	15.0	2.0	2.5	11.2
Control methemoglobin levels (%)	4.2	3.5	---	---	3.5	---

McCord et al., 1941 (Ref. 236).

Chronic

Kellerman, 1956 (Ref. 242), noted that experiments on white rats to date had offered no cause for concern in welders exposed to fumes from basic covered electrodes. Lehmann, 1956 (Ref. 242), reported a series of experiments in which 18 rats (strain, sex, and age not provided) were exposed for 6 hours a day (number of days not specified) to fumes from 6 to 7 kg of FOX EV 50 electrodes (German manufacturer), with an average fume concentration of 34 mg/m^3. Iron, calcium, manganese, and alkalies were the main dust constituents. Gaseous fluorine compounds could not be demonstrated; carbon monoxide levels averaged 120 ppm. With the 34 mg/m^3 exposure series, no unusual behavior or body weight change were noted when compared to the controls; blood counts remained normal. Bleeding from the nasal membranes occurred in younger rats and, to a lesser extent, in the controls. Reproduction was not affected. Three of 18 rats died; no controls died. Cause of death was seemingly unrelated to exposure to welding fumes. All of the experimental animals showed inflamed gastric and intestinal mucous membranes attributed to the chronic exposure. Of 18 rats exposed to approximately 300 mg/m^3 of welding fumes from similar electrodes, 15 died. Severe inflammatory bronchitis probably caused death. Marked iron pigment deposition in the lungs was noted, but there was no detected abnormality in trachea, heart, spleen, liver, kidneys, brain, or spinal bone marrow. These animals died before chronic gastric mucosal inflammation could develop (the duration of exposure until death was not specified).

Erman and Rappoport, 1970 (Ref. 19), reported the toxicity of welding fumes and gases in albino rats (strain, sex, and age not given) exposed to fume levels of 150 to 180 mg/m^3, 122 to 140 ppm CO, 0.5 to 0.6 volume percent CO_2, 3.6 to 6.8 mg/m^3 nitrogen oxides (1.92 to 3.62 ppm based on NO_2), and up to 0.18 ppm of ozone. The welding fumes were produced from CO_2 shielded arc welding with bare iron electrodes containing up to 2.1 percent manganese.

Three groups of 10 rats each were exposed for 4 hours daily, 6 days per week, for 1, 3, and 6 months. Thirty control rats were used.

The gain in body weight of the control animals over the 6-month period was twice that of the experimental rats (actual body weights were not given). The relative weights of liver and kidney (as percentages of body weight) of rats exposed to welding aerosol were significantly higher than those of the control rats (p values not given), with the exception of the kidney weights after 3 and 6 months of exposure. Relative kidney and liver weights are presented below:

Exposure period (months)	Relative wt. of liver (% body weight)		
	Control	Experimental	% change
1	2.73	3.30	+21
3	2.80	3.22	+15
6	2.94	3.50	+17.8

Exposure period (months)	Relative wt. of kidney (% body weight)		
	Control	Experimental	% change
1	0.32	0.39	+21.8
3	0.32	0.34	+ 6.2
6	0.34	0.36	+ 5.8

Pathologic findings at autopsy of these rats included gross venous congestion in the brain, lungs, liver, and kidneys after 3 or 6 months of exposure. Small, confluent, brown pigmented areas were noted in lung and subpleural tissue. After 3 months of exposure, a considerable number of dust-cell foci were present in the lungs. In areas free of dust-cell foci, the interalveolar walls were thickened with argyrophil (reticulin) fibers. These conditions were more pronounced after 6 months of exposure to the welding fumes. Degenerative changes were also found in the central nervous system, heart, liver, kidneys, and ovaries of animals exposed for 6 months. The degenerative effects of exposure to welding fumes and gases were qualitatively more severe than the lung reactions, characterized by the authors as weak focal fibrosis. (Nervous system abnormalities are discussed in Chapter 3.).

Felczak, 1967 (Ref. 300), studied the development of lung changes related to the inhalation of welding fumes and gases in 41 mature Wistar rats of both sexes. The rats were divided into 4 groups. Rats in Group 1 were exposed to welding fumes (no fume concentrations stated) from rutile covered electrodes for 95 days (240 hours). Rats in Group 2 were exposed to welding fumes and gases from rutile and basic covered electrodes for 110 days (356 hours) at a shipyard welding site. Rats in Group 3 were exposed under identical conditions as in Group 2, allowing a recovery period of 30 days after exposure to fumes. Rats in Group 4 were exposed at the same site for 139 days (576 hours).

The rats were x-rayed before and after exposure, and lung tissue was examined for iron deposition, collagen, and elastic fiber content. Results showed that deposits in lung tissue consisted mainly of iron particles with small amounts of manganese and silica. Inflammation of interstitial tissue was noted and ascribed to the irritating action of the welding gases and other oxides, rather than to the presence of iron deposits (Ref. 300).

Guskova and Komovnikov, 1974 (Ref. 20), exposed 100 white rats (strain, sex, and age not given) to inhalation of welding fumes from two types of rutile covered electrodes, OZS-4 and TsL-11 (compositions not given), for 3 hours daily, 5 days a week, for up to one year. Fume levels of 290 to 310 mg/m^3 were maintained. Animals were killed at 3 month intervals during the exposure. The number of alveolar phagocytes and the phagocytic index were observed to increase shortly after exposure, gradually returning to normal after 12 months. Numbers of destroyed dust-laden cells were also elevated shortly after treatment with OZS-4 fumes, but not TsL-11 electrode fumes. The collagen content (method of measurement not specified) of the lungs was reported in order to evaluate fibrotic changes; the collagen content increased as the length of exposure increased, as shown in Table D3. The authors concluded that the fume particles stimulated phagocytosis and were of relatively low fibrogenicity in the lungs of rats.

Table D3
Collagen content of lungs of rats exposed to welding fumes

		Collagen (mg) in exposed rat lungs	
Exposure (months)	Collagen (mg) in controls	Fumes from OZS-4 electrodes	Fumes from TsL-11 electrodes
3	9.1	10.1	12.2
6	9.7	17.8	15.6
9	10.4	19.8	16.9
12	12.4	20.4	19.6

Guskova and Komovnikov, 1974 (Ref. 20).

Samoilova and Kireev, 1975 (Ref. 21), studied inhalation effects of welding fumes generated from electrodes containing CuO, MnO_2, Fe_2O_3, HF, and TiO_2 (I) or NiO (II) in 60 white rats exposed for 1 to 12 months. Twenty control rats were also examined (strain, sex, and age of rats not given; fume levels were 40 to 60 mg/m^3). Desquamation of the lung tissue, interstitial deposition of metal oxides, bronchitis, and emphysema were observed to develop in the rats (occurrence rates not given). Welding fumes containing V_2O_5, MnO_2, Fe_2O_3, and HF (III) also caused fibrosis, pneumonoconiosis, and sclerosis around pulmonary blood vessels and bronchi in an unspecified percentage of 30 exposed white rats (1 to 12 months; 40 to 60 mg/m^3). The fume compositions are given below.

	Fume concentrations (g/kg of electrode consumed)		
	I	II	III
CuO	5.2	6.3	–
MnO_2	0.55	2.62	0.51
Fe_2O_3	3.1	2.73	4.4
TiO_2	0.28	–	–
HF	2.8	1.5	2.2
NiO	–	2.7	–
V_2O_5	–	–	0.63

The effects on rats of welding fumes from nickel-chromium electrodes used for stainless steel welding were reported by Migai and Norkin, 1965 (Ref. 234). The animals (28 males, 26 females, 150 g; 20 controls) were exposed to welding fumes from electrode 606/11 for 5 to 6 hours daily, 6 days a week, for 9 months. Compositions of the welding fumes in mg/m^3 in the exposure chamber were MnO_2-0.57, CrO_3-1.45, F-4.45; gases included 4.1 ppm NO, NO_2-trace, and 3.7 ppm HF.

The result of blood analysis showed no significant difference in hemoglobin content, red blood cell counts, and white blood cell counts between the control and experimental groups. Oxygen uptake for rats exposed to welding fumes was higher than for the control group after 3 months, but lower than for the controls after 6 or 9 months of exposure. The experimental group also showed a slower rate of body weight gain and was less sensitive to electrical stimuli as compared to the controls, as shown below:

	Experimental rats	Control rats
Oxygen uptake (ml/hr/kg)		
3 months exposure	2490	1905
6 months exposure	1678	1702
9 months exposure	1285	1922
Skin sensitivity (volts)		
3 months exposure	18.6	20.5
6 months exposure	25.2	22.2
9 months exposure	25.7	22.3

Chronic Effects of Copper, Nickel, Aluminum, and Magnesium Fumes

Vlasova-Pryadilova, 1971 (Ref. 22), noted the chronic effects of welding fumes from copper welding and cutting using copper electrodes or "Komsomolets-100" (K-100) electrodes on the respiration rate, dry weight, and collagen content of the lungs of rats. White rats exposed to 120 to 150 mg/m^3 of welding fumes for 1, 3, 6, or 10 months (hours per day not specified) and then killed, developed catarrhal desquamative changes in alveoles, interstitial peribronchial pneumonia, and pneumosclerosis. The coniotic process differed from silicosis and was more pronounced when K-100 electrodes were used. Changes in the lungs are listed in Table D4.

The author felt that copper oxide lung toxicity was due to a blocking of sulfhydryl protein groups by copper, resulting in acute irritation and pneumonia, and leading to the chronic fibrotic changes noted above.

Vorontsova et al., 1969 (Ref. 239), exposed albino rats (sex and age not specified) to welding fumes produced by burning either copper wire or electrode K-100 for up to 10 months in order to study pathological changes in the respiratory organs. The coating of electrode K-100 contained feldspar, fluorspar, copper, silicate, and ferromanganese (average total dust, 20.8 g/kg; average manganese oxide, 0.28 g/kg). Unfortunately, the length of the daily exposure was not reported, and the concentration of any gases produced was not measured.

After 2 months of exposure to 50 to 80 mg/m^3 of welding fumes from K-100 electrodes, one of the 40 rats died. There was 35 percent mortailty among the 40 rats exposed to copper wire fumes for 6 months, compared to 25 percent mortality after exposure to electrode K-100 fumes.

The animals died from hemorrhagic pneumonia or diffuse multifocal bronchopneumonia, occasionally purulent. Histological examination of the lungs and other organs of rats after 1, 3, 6, or 10 months of exposure showed progressive inflammation of bronchi and interstitial tissue of the lungs with emphysematous foci, lymphoid follicular hyperplasia, and diffuse fibrosis. The changes were more pronounced in animals exposed to copper oxide from copper wire than to welding fumes from electrode K-100.

Histological examination of the lung tissue of rats after inhalation of welding fumes containing copper, nickel, and iron oxides was reported by Arutyunov et al., 1976 (Ref. 23). They exposed 1125 mongrel albino rats (340 per group; sex, age, and body weight not given) to welding fume levels of 70 to 80 mg/m^3, containing copper oxide, copper, and nickel oxides or iron oxide, 3 hours per day, for 4 to 9 months. There were an additional 102 controls. The rats were killed 5, 15, 30, 45, 60, 120, and 240 minutes from the beginning of the inhalation; 3, 6, 9, 15, 21, 28, and 42 hours; 3, 5, 7, 14, 21, 30, and 45 days; and 2, 3, 4, 6, and 9 months and up to 3 years after the termination of the experiments.

Pathological examination of the lung tissue showed that rats exposed to welding fumes containing copper or copper and nickel oxides developed pneumosclerosis, and emphysema after 4 to 6 months of exposure. Early accumulation of particulates and lipoprotein in alveoles progressed to pulmonary alveolar proteinosis (PAP), a chronic condition reflecting a prolonged functional lung stress (for example, from chronic hypoxia), which was not observed in controls or rats exposed to iron oxide fumes. PAP disappeared 2 to 3 years after termination of the experiment. Emphysema and fibrosis, however, would be permanent lung changes.

Likhachev et al., 1975 (Ref. 241), exposed 500 rats (sex, strain, and age not presented) to arc welding fumes containing copper oxides, killing the animals at varying intervals (3 days to 2 years) from the start of the exposure, which only lasted 9 months for 3 hours per day. Pulmonary alveolar proteinosis (lipoproteinosis), fibrosis, and delayed macrophage replacement were noted, similar to the previous study (Ref. 23). The changes were proportional to fume concentration and the duration of ex-

Table D4
Lung changes in rats exposed to welding fumes from copper-containing electrodes

Condition	Group	Baseline data	Exposure period (months) 1	3	6	10
Respiration rate (per minute)	1	101	174	232	210	204
	2	106	186	210	225	210
	Control	105	101	108	108	107
Dry weight of lungs (mg)	1	---	908	1333	1850	2070
	2	---	623	952	1696	1741
	Control	---	148	198	181	338
Collagen content of lungs (mg)	1	---	26.1	40.9	70.1	87.2
	2	---	13.6	39.1	64.3	86.0
	Control	---	2.8	2.2	3.2	6.1

Vlasova-Pryadilova, 1971 (Ref. 22).

posure. Fume concentrations were not stated, however.

The effects on white rats of fumes produced by argon metal arc welding with aluminum and aluminum-magnesium alloy electrodes were reported by Leonicheva, 1965 (Ref. 240). A preliminary test with white rats showed that inhalation for 1 year of the welding fumes generated from Al-Mg alloy electrodes produced low mortality, although the animals died later from pneumonia, bronchitis, and pulmonary abscesses.

Repeating the inhalation experiment, 40 rats were exposed to welding fumes from Al electrodes, while 40 rats inhaled fumes from Al-Mg electrodes, and 20 served as controls. The fume concentration range was 120 to 140 mg/m^3; exposure durations were a few days to 12 months, 3 hours daily. Pathological changes observed in the two experimental groups were similar. Dust deposits were found in pulmonary tissue; they were partially eliminated via the respiratory passages and lymphatics. Precollagen and collagen fibers were found around bronchi and blood vessels. These fibrotic changes were more pronounced in animals exposed to Al-Mg dust than to Al dust alone.

Central Nervous System Effects

Erman and Rappoport, 1970 (Ref. 19), noted central nervous system abnormalities in albino rats (strain, sex, and age not given) exposed to welding fumes and gases from CO_2 shielded arc welding with bare steel electrodes containing up to 2.1 percent manganese. Fume levels averaged 150 to 180 mg/m^3, with 120 to 140 ppm CO, 0.5 to 0.6 volume percent CO_2, 3.6 to 6.8 mg/m^3 nitrogen oxides (1.9 to 3.6 ppm as NO_2), and up to 0.02 ppm of ozone. Three groups of 10 rats each (30 controls) were exposed for 4 hours daily, 6 days per week, for 1, 3, and 6 months.

Table D5 lists these electrophysiological changes in rheobase, chronaxy, and lability. Rheobase is defined as the minimum potential of electric current necessary to produce nerve stimulation, determined with a 10 msec pulse duration. Chronaxy is the minimum time an electric current must flow at a voltage twice the rheobase to cause muscle contraction. The lability index of the caudal neuromuscular apparatus is the number of pulses per second at which the transition from clonic to tonic spasm occurs. Differences between initial index and values after exposure were stated to be significant (p level not given), with the exception of chronaxy after 1 and 3 months of exposure.

Cholinesterase activity alterations were not noted in these exposed rats. Levels of acetylcholine in various organs are listed below, in μg acetylcholine per minute per g tissue.

Tissue	Controls	3-month exposure	6-month exposure
Heart	223	236 ($p = 0.5$)	159 ($p < 0.05$)
Lungs	132	207 ($p < 0.01$)	78 ($p = 0.001$)
Liver	190	235 ($p = 0.05$)	183 ($p > 0.20$)
Kidneys	71	150 ($p < 0.001$)	110 ($p = 0.05$)
Spleen	228	260 ($p = 0.05$)	233 ($p = 0.50$)
Cerebral cortex	989	1005 ($p > 0.10$)	877 ($p = 0.05$)
Medulla oblongata	950	1146 ($p < 0.01$)	883 ($p > 0.10$)
Erythrocytes	88	>116 –	86 ($p > 0.20$)
Serum	74	>116 –	76 ($p > 0.20$)

Examination of the brains of exposed rats revealed degenerative changes in the middle and deep layers of the cerebral cortex and subcortex after 3 months of exposure. Dystrophic changes were more pronounced in cortical neurons and subcortical region in rats with longer exposures.

Liver Effects

The effect of the inhalation of welding fumes on oxygen consumption by liver tissue of albino rats was studied by Byczkowski et al., 1965 (Ref. 245). Rats (160 to 200 g) were exposed to welding fumes containing oxides of iron, silicon, manganese, titanium, clay, nitrogen oxides, and ozone from EP47-28 covered electrodes for up to 15 weeks, 3 hours daily, for 6 days per week. At various intervals during the experiment and also during a 40-day recovery period, rats were killed and their livers homogenized in 0.15 M KCl. Oxygen uptake

Table D5
Electrophysiological indices in albino rats exposed to welding fumes and gases

	Rheobase		Chronaxy		Lability	
	voltage	% change	msec	% change	pulse/sec	% change
Initial index	2.35		0.037		12	
1 mo. exposure	2.55	+ 8.5	0.036	-2.7	13	+ 8.3
3 mo. exposure	2.66	+13.5	0.034	-8.1	14	+16.6
Initial index	2.00		0.06		12	
6 mo. exposure	1.68	- 16.0	0.022	-56	15	+23.8

Erman and Rappoport, 1970 (Ref. 19).

of the homogenates was measured by Warburg manometry. Great variation in the oxygen consumption was obtained, but no significant deviations from the control values could be observed. Table D6 notes the oxygen consumption values obtained.

Table D6
Oxygen consumption in homogenized liver tissue from rats exposed to welding fumes

	O_2 consumption ($\mu g\ O_2$/mg N)
Control	av. 8.9389
15 weeks exposure	17.9179
3 days recovery	11.5777
6 days recovery	3.3615
26 days recovery	26.5643
30 days recovery	15.9029
40 days recovery	9.9583

Byczkowski et al., 1965 (Ref. 245).

Reproductive System Effects

The studies that follow employed fumes from EP 47-28, a rutile covered Polish electrode. Only rats have been examined following exposure to these fumes. While it appears that exposure decreased numbers of pregnant females, litter size, and fetal weight; reduced fertility of exposed male rats; and caused histopathological changes in the male and female reproductive organs, further research is necessary, with fumes from widely used electrodes and a greater variety of animal species, in order to appropriately evaluate such findings.

The effect of welding fumes on reproduction in rats was studied by Dabrowski et al., 1966 (Refs. 243 and 244). Twenty-five rutile covered electrodes (Polish EP 47-28) were consumed 3 hours daily to generate welding fumes (222 mg/m^3) for the animal exposure chamber (7 m$_3$). Fume composition in the chamber during exposure was 102 mg/m^3 iron, 15.2 mg/m^3 silicon, 9.2 mg/m^3 manganese, 3.53 mg/m^3 titanium oxide, and 8 mg/m^3 nitrogen oxides (4.3 ppm as NO_2).

A total of 75 mature female Wistar rats, 3 to 4 months old, 145 to 185 g, were divided into 5 groups of 15 each. Rats in Groups 1, 2, and 3 were exposed to welding fumes for 32, 82, and 102 days, respectively; those in Group 4 were exposed for 80 days, followed by a rest period of 102 days. Group 5 served as a control. Rats in Groups 1, 2, and 3 were mated with unexposed males for 72 hours during the exposure period, and those in Group 4 were mated after the 80-day exposure (mating dates not specified).

Toward the end of gestation, female rats were killed and examined. The number of rats that became pregnant, the average number of live fetuses per litter, the average fetal weight, and malformations are listed as follows.

Group	1	2	3	4	5
Exposure (days)	32	82	102	80 + 102 day rest	Control
No. of pregnant rats	11/15	6/15	2/15	0/15	13/15
Average no. of live fetuses per litter	9	8	6.7	–	9
Average fetal weight (g)	82	62	54	–	89
No. of fetal malformations	none	none	not given	–	none

Pathological examinations showed aggregation of phagocytes that contained inorganic iron deposits in the lungs after 32 days of exposure. Similar cells with siderotic cytoplasmic granules were detected in the uterus, ovaries, decidua (mucous membrane of the pregnant uterus), and placenta of rats exposed to welding fumes for 82 and 102 days (Ref. 243).

To study the effects of welding fumes on male rats, Dabrowski et al., 1966 (Ref. 244), exposed 2 groups of mature Wistar rats (6 to each group) to similar fume conditions as those for the aforementioned female rats (Ref. 243). One group of male rats was exposed for 100 days, and immediately thereafter mated with unexposed females (5 females to 2 males for 72 hours). None of the female rats became pregnant. Another group of male rats was exposed for 100 days, followed by a recovery period of 80 days, and then mated with unexposed females. Only one quarter of the females (4 of 16) became pregnant. The average fetal weight was 86 g (89 g in controls) and there were an average of 8 live fetuses per litter (9 in controls). No fetal malformations were detected, and no postnatal mortality was observed in 6 months.

Histological examination of the testes of rats exposed for 102 days showed edema in the interstitial tissue and aggregation of cells containing siderotic granules. In the second group of rats, additional degenerative changes in the testes were observed, such as desquamation and degeneration of germinal epithelial cells (noted in the lumina of the seminal ducts), cast formation of the degenerated germ cells, caseous necrosis in the seminal ducts, deposits that resembled the necrotic calcified tissue in the lumina of the seminal ducts, and absence of spermatogenic elements in the seminal ducts (Ref. 244).

Effects in Rabbits

Effects noted in the literature review in this section include reduced rate of weight gain (Refs. 192 and 236), varying degrees of lung damage from inflammation, edema and hemorrhage (Refs. 27 and 246), and methemoglobinemia (Ref. 236).

Von Haam and Groom, 1941 (Ref. 235), noted that

rabbits seem more resistant to toxic effects of inhalation of welding fumes and gases than rats, guinea pigs, or mice. This species specificity may be important for future animal experimentation.

Investigations of rabbits are presented below, according to acute or chronic duration of exposure.

Acute

Titus et al., 1935 (Ref. 246), exposed 16 rabbits (sex, strain, and age not provided) to various concentrations of welding fumes and gases for 48 to 510 minutes. Fumes were generated by the dc cutting of blue annealed steel (SAE 10-15) using bare iron electrodes; ferric oxide levels were 35 to 250 mg/m^3. One to fourteen percent CO_2 was added to the air in the exposure chamber in order to increase the rabbits' respiratory rate. Twelve rabbits were exposed to filtered welding fumes, in which particulates were removed by 6-ply Air Mat paper (a material used for cleaning ventilators). Eight rabbits were exposed to pure commercial ferric oxide powder, which was blown into the exposure chamber by compressed air. The results of these experiments are presented

Table D7
Lung pathology of rabbits exposed to welding fumes, gases, and ferric oxide powder

Number of rabbits	Av. conc. of Fe_2O_3 (mg/m^3)	Exposure time (min.)	CO_2 conc. (%)	Results
2	>90	48	>3.4	1 normal: 1 with inflamed trachea (examined 4 hrs. later)
1	63	157	<3.4	Appeared normal, slight pulmonary edema
1	49	224	<3.4	Gross edema, inflamed trachea, hemorrhagic lungs
2	39	312	<3.4	1 normal; 1 with edema, lungs hemorrhagic 1 day later
3	175	360	3-14	Abnormal appearance, 2 with edema; 1 with inflamed lungs
1	250	241	1-3	Died; lungs normal
1	250	364	1-7	Died; lungs edematous with brown pigmentation
1	175	510	1-7	Slight edema, lungs mottled, trachea inflamed
1	---	86	4-9	Normal
1	210	240	4-9	Lungs mottled
1	175	323	4-9	Bronchioles edematous, lungs abnormal after 2 days
1	175	323	4-9	Slight edema
5	35-100 Iron oxide powder	39-291	$\sim$5	All normal, taken out at intervals; 1 autopsied next day
3	190 Iron oxide powder	255	<6	Normal; tracheas slightly inflamed
3	Filtered fumes	322	3-10	Normal, slight edema in two
3	Filtered fumes	360	3-14	Abnormal appearance; 2 edematous; 1 edematous next day
3	Filtered fumes	510	1-7	Normal
3	Filtered fumes	313	4-9	Lungs of all 3 mottled; 2 died, both edematous

Titus et al., 1935 (Ref. 246).

in Table D7.

Because iron oxide powder did not cause any notable adverse effect, but both filtered and unfiltered welding fumes caused pulmonary edema, hemorrhage, and pigmentation to occur, the authors concluded that the gases (ozone and nitrogen oxides) were solely responsible for the lung damage in these rabbits.

Von Haam and Groom, 1941 (Ref. 235), exposed 2 rabbits (sex, strain, and age not provided) to 1600 to 2600 mg/m^3 of welding fumes from A5 iron-based covered electrodes (17.6% Fe_2O_3, 9.5% Cr_2O_3, 10.6% Mn_3O_4, 16.8% CaO, 16.1% Na_2O, 5% Al_2O_3, 14.6% F) for 6 hours. The animals showed no abnormal lung reaction. A rabbit exposed for 6 hours developed bronchial ulceration. Repeated exposures, 1/2-hour twice a day for 2 months, produced no reactions in the rabbits. The authors concluded that rabbits were more resistant to welding fume toxicity than other animals tested under similar conditions of exposure, including rats, guinea pigs, and mice.

To study the physiological effects of fumes and gases produced by arc welding with bare, washed iron electrodes, Harrold et al., 1940 (Ref. 192), exposed a total of 58 rabbits (44 controls; sex, age, and strain not given). Various voltages (27 to 44 volts) were used during the welding. The animals were exposed for 6 hours daily, 5 days per week, for 120 to 228 hours (20 to 38 days). Analysis of the composition of the particulate fraction of the welding products showed iron oxide to be the major component (35 to 398 mg/m^3, varied with voltage across the arc). The content of manganese varied from 0.2 to 3.3 mg/m^3. The gaseous products consisted mainly of nitrogen oxides (reported as NO_2), which varied with the voltage (29 ppm at 27 volts to 70 ppm at 44 volts). The level of ozone was high near the arc (10 to 32 ppm within 1 in. of the arc) and decreased with increasing distance from the arc (0.2 to 1 ppm at center of room). Carbon monoxide, chlorine, and other gases were not detected. The percentages of oxygen and carbon dioxide were considered normal for room air.

Over 90 percent of the animals survived the exposure. Body weight gain of the experimental animals was slightly less than that of the controls. Autopsies carried out at various intervals during the experiment showed pulmonary deposition of iron particles that migrated to the periphery of the lung and the tracheobronchial lymph nodes. No edema of the lung or any portion of the respiratory tract was detected. X-ray examination of the exposed animals and the extirpated lungs showed no abnormal findings.

McCord et al., 1941 (Ref. 236), investigated the toxicity of fumes and gases from covered arc welding electrodes in rabbits. Formation of methemoglobin and reduced rate of weight gain were two important results of the experiments. No significant lung changes were noted. Analysis of the coating of the welding rod and the fumes generated during welding showed the following compositions.

	Rod coating (%)	Airborne fume (%)
Fe_2O_3	8.8	79.0
MnO_2	10.9	5.0
TiO_2	41.5	5.4
SiO_2	20.5	8.4
CaO	2.5	–
MgO	5.4	–
Organic matter, moisture	6.9	1.7

Animals were exposed to fumes and gases from coated electrodes, welded in a 28.3 m^3 chamber. Welding was performed at 44 V and 300 to 350 A. Evolved gases included more than 20% O_2, 0.22% CO_2, and 0.16 to 2.07 ppm ozone. Animal exposures averaged 24 ppm NO_2, 444 mg/m^3 Fe_2O_3, and 15.5 mg/m^3 Mn. Traces of titanium and calcium oxides were noted. Silica was found in quantities averaging 61 mg/m^3 as SiO_2.

Groups of 16 male and female rabbits (strain and age not provided) were exposed to welding fumes and gases for 6 hours daily (welding at 1/2-hour intervals), 5 days per week, for 45 days (65 calender days).

All rabbits survived, except one control. The average weight gain for exposed rabbits was 1 kg, compared with 1.5 kg for controls. X-ray examination of the animals before and after the exposure period showed no exposure-related abnormalities. Complete blood analysis (cell counts and hemoglobin determinations) every 7 to 14 days during the exposure revealed no significant differences between the exposed and the control animals. However, substantial quantities of methemoglobin were found in the blood of the exposed rabbits. Shortly after exposure, the level of methemoglobin returned to the normal as shown in Table D8.

Table D8
Methemoglobin levels in rabbits exposed to welding fumes and gases

	Male			Female		
Exposure (days)	6	14	27	6	14	45
Average methemoglobin levels (%)	2.8	1.5	2.7	0.6	0.2	2.9
Control levels	---	0.6	---	---	0.6	---

McCord et al., 1941 (Ref. 236).

Chronic

Garnuszewski and Dobryznski, 1966 (Ref. 27), exposed 10 rabbits (sex, strain, and age not provided) to welding fumes from electrode EP 47-28P (23% Fe_2O_3, 8% SiO_2, 14% TiO_2, 9% MnO_2, 2% Na; Al, Ca, Ni, V, Cu $<$ 1%) for 4 hours daily for 180 days. Fume levels were not provided; the authors stated that levels were similar to those found in shipyard welding environments. The lungs of the 10 rabbits killed after 6 months of exposure showed slight septal thickening but not fibrotic nodules. They also exposed guinea pigs to similar welding fume levels; siderosis and silicosis reactions developed, and deaths occurred from bronchial pneumonia. (The details of the guinea pig experiments are presented in the guinea pig section.)

Effects in Guinea Pigs

Studies include lethal and toxic results of guinea pig exposures to welding fumes and gases. One experiment revealed that guinea pigs were most sensitive to the inhalation effects, when compared to rats, rabbits, and mice (Ref. 247), and further found the particulates, but not the gases from welding with basic covered electrodes, to be the toxic substances (Ref. 247). Pulmonary edema, bronchial pneumonia and pleural exudation, siderosis, and silicosis were observed to develop occasionally in exposed guinea pigs (Refs. 235 and 246), although lung reactions usually were absent (Refs. 235 and 248).

These animal experiments are reviewed below, by acute or chronic duration of exposure.

Acute

The toxic effects of welding fumes from basic electrodes on guinea pigs were investigated by Kawada et al., 1964 (Ref. 247). Three types of electrodes were used for these studies: an ilmenite ($FeTiO_3$) covered electrode, B-17, (corresponds to the British BS E-316 or E-416 electrodes or the German DIN-ES, Erzsaurer Typ electrode); and two basic covered electrodes, LB-52, and LBM-52. The compositions of the coating material of the B-17 and LB-52 electrodes are shown in Table D9.

Table D9
Analyses of coating materials of electrodes B-17 and LB-52

	Electrode type	
Constituent	B-17 (ilmenite) (%)	LB-52 (basic) (%)
SiO_2	24.11	20.84
FeO	24.39	8.17
TiO_2	11.67	4.77
MnO	19.41	6.02
CaO	7.30	32.42
F_2	---	7.70
Alkali	2.23	4.03
Organic substances	5.46	---
CO_2	5.78	20.11
Al_2O_3, MgO	---	trace

The compositions of the soluble and insoluble constituents of the welding dusts from these electrodes are given in Table D10.

Analyses of the gaseous products from the welding with B-17 and LB-52 electrodes revealed the following.

	O_2 (Vol. %)	CO_2 (Vol. %)	CO (Vol. %)	NO_2 (ppm)
B-17	15.6	4.3	trace	<10
LB-52	17.6-19.2	3.0-3.8	0.01	17-18

Welding was performed on a steel plate in a steel box (13 x 13 x 41 cm), and the fumes and gases were introduced at a constant rate of 5 liters/min into a glass desiccator in which the animal was kept for inhalation tests. After burning the 4 x 400 mm electrode for one minute (170 A, ac), the welding was stopped, and the desiccator vents were simultaneously opened. The animal was taken out of the desiccator after one hour of exposure and observed for several days.

A preliminary test was conducted with a mature male mouse, a white rat, a rabbit, and a guinea pig to determine the most sensitive animal for the inhalation test. The welding fumes and gases from the basic covered electrode, LB-52, were found to be lethal only to the guinea pig. Accordingly, this animal was used for all subsequent inhalation experiments.

Exposures revealed the fumes and gases generated by burning basic covered LB-52 electrodes to be much more toxic to guinea pigs than those produced by burning ilmenite covered B-17 electrodes. Toxicity of the fumes from the improved basic covered electrode, LBM-52, ranked in between. After exposure to the fumes and gases from burning LB-52 electrodes, most of the guinea pigs survived only for several minutes, and 10 out of 12 animals died within 24 hours. Similar exposure to the welding fumes and gases of the ilmenite covered B17 electrode killed only 2 out of 10 guinea pigs.

Histopathological examination of the respiratory tract and lungs of animals that died after exposure to the welding fumes and gases showed pulmonary emphysema, blood stasis, dust deposition, and sometimes bleeding, bronchopneumonia, lung collapse, and pulmonary edema. Blood methemoglobin levels were higher (3 to 9 mg/ml) for animals exposed to welding fumes and gases from basic covered LB-52 electrodes than either those exposed to ilmenite covered B-17 electrodes or the control animals (0 to 2 mg/ml).

Electronmicrograms of the lung tissues of guinea pigs exposed to welding fumes from electrodes B-17 or LB-52 showed particle-containing phagocytes in the alveolar lumen and the alveolar wall. No particles were found in the alveolar epithelium or in the capillary endothelium (Ref. 247).

In order to find out whether the particulate matter or gaseous products in the welding fumes from the covered electrode LB-52 were responsible for the toxic effects, nine guinea pigs were exposed to welding gases filtered through an absorbent cotton filter to remove all particulates for one hour and observed for several days.

Table D10
Analyses of welding fumes from electrodes B-17, LB-52, and LBM-52[a]

Electrode	B-17		LB-52		LBM-52[a]	
Constituent	Insoluble (%)	Soluble[b] (%)	Insoluble (%)	Soluble (%)	Insoluble (%)	Soluble (%)
SiO_2	16.30	0.15	6.20	0.10	6.30	0.05
Al_2O_3	trace[c]	trace	trace	trace	trace	trace
Fe_2O_3	54.41	0.98	26.55	0.15	30.34	0.10
MnO	12.61	0.05	4.03	trace	4.85	trace
TiO_2	1.94	trace	0.60	trace	0.18	trace
CaO	1.19	trace	12.76	trace	11.20	trace
MgO	0.15	trace	0.15	trace	3.40	trace
CO_2	0.18	0.20	0.17	0.50	0.15	0.50
BaO	---	---	3.20	trace	3.55	trace
Na_2O	trace	4.87	trace	10.17	trace	21.77
K_2O	trace	6.03	trace	16.70	trace	3.38
F^-	---	---	5.65	12.52	5.40	11.38
Total	87.78	12.28	59.31	40.34	65.27	37.18
Sum total	99.06		99.65		102.45	

a. Results with this electrode are discussed elsewhere.
b. Soluble in boiling water in 5 minutes.
c. Less than 0.05%.

None of the animals died, and all appeared normal. The composition of the filtered gaseous products was: O_2, 18.4 volume percent; CO_2, 3.8 volume percent; CO, trace; and NO_2, 25 to 60 ppm. Thus, the particulate matter in the welding fumes was considered to be responsible for the toxic effects (Ref. 301).

Von Haam and Groom, 1941 (Ref. 235), exposed 3 guinea pigs (age and sex not provided) for 6 hours to 1600 to 2600 mg/m^3 of welding fumes from A5 electrodes (17.6% Fe_2O_3, 9.5% Cr_2O_3, 10.6% Mn_3O_4, 16.8% CaO, 16.1% Na_2O, 5% Al_2O_3, 14.6% F). Pulmonary edema and respiratory tract irritation developed, One guinea pig died from acute pulmonary edema. Three of 5 guinea pigs exposed to welding fumes for 1/2-hour, twice daily, for 2 months died of severe pulmonary reactions. None of 3 guinea pigs developed any reaction to filtered welding gases (exposure time not provided; gases contained 19.5% O_2, 78% N_2, 0.7% CO_2, trace CO, and qualitative confirmation of presence of nitrogen oxides). The authors mentioned that fume levels of 800 to 1600 mg/m^3 from similar electrodes (F5 and F9) caused deaths in some exposed guinea pigs, but experimental details were not included.

Chronic

Garnuszewski and Dobrzynski, 1966 (Ref. 27), studied the histological lung changes in guinea pigs exposed to welding fumes from electrodes containing various quantities of silica and metallic oxides.

Seventy-two guinea pigs (10 controls) were exposed to welding fumes from electrode EP52-28P (13% MnO, 26% SiO_2, 6% Al_2O_3, 18% Fe_2O_3, 9% TiO_2, 7% CaO, 2% MgO, 6% CO_2) for 110 days, 4 hours daily (Expt. 1). Forty guinea pigs (10 guinea pigs as controls) were exposed to welding fumes from electrode EP47-28P (23% Fe_2O_3, 8% SiO_2, 14% TiO_2, 9% MnO_2, 2% Na; Al, Ca, Ni, V, Cu $<$ 1%), for 4 hours daily for 180 days (Expt. 2). Fume and dust concentrations were stated to be similar to those found in shipyard welding environments. Thirty guinea pigs exposed to welding fumes from electrode EP52-28P died before the end of Expt. 1. The remaining animals were killed at intervals and examined histologically.

Autopsy of those animals that died during the experiment showed hyperemia and edema of lungs, numerous suppurating foci of bronchial pneumonia, and exudates in the pleural cavity in some cases. Lungs of guinea pigs that were killed on the 75th and 110th days of exposure to fumes from electrode EP52-28P contained gross bluish-gray colored subpleural accumulations of dust-laden phagocytes. Microscopic examination of lungs revealed diffuse interstitial pneumonia, with iron and silica-containing cells in thickened alveolar septae, alveoli, and bronchioles. During exposures to fumes from electrode EP47-28P, 25 out of 40 guinea pigs died, 13 in the initial 4 weeks. Ten were killed after 6 months, 5 after 7 months, and the rest 10 months after the end of the exposure. Lung histopathology was similar to that for animals in Expt. 1, consisting of alveolar wall damage, numerous dust-laden phagocytes containing iron and silica in alveolis and bronchi, and small fibrotic nodules in the thickened alveolar septae.

In conclusion, fumes from EP52-28P electrodes caused siderosis and silicosis in guinea pig lungs; fumes from EP47-28P lead to siderosis under the same experimental conditions (Ref. 27).

Gadzikiewicz and Dominiczak, 1968 (Ref. 248), exposed 50 guinea pigs (500 g) to welding fumes containing 1.72 to 2.52 mg/m^3 manganese, 4.33 to 6.08 mg/m^3 silica, and 21.44 to 24.83 mg/m^3 iron from covered 5 mm diameter electrodes (EP47-28P) for 180 days. Twenty-five controls were also used. The exposure schedule was not further specified. Examination during the exposure failed to reveal pathological changes or weight loss in the guinea pigs. Following the 180 days of exposure, animals were killed. The authors detected no abnormalities in gross examinations or histological preparations of liver, kidneys, adrenals, spleen, or parenchymatous organs. Similar results were obtained during and after guinea pig exposures to welding fumes from EP52-28P 5 mm diameter covered electrodes; the fumes contained 1.44 to 2.52 mg/m^3 manganese, 1.87 to 4.12 mg/m^3 silica, and 4.38 to 9.65 mg/m^3 iron. Note that the iron content of fumes in the latter exposure was almost 4 times lower than in the former with EP47-28P electrodes, although the effects of the two exposures were equally negative.

Effects in Mice

Two studies are detailed below: one regarding the general toxicity of fumes and gases from CO_2 shielded arc welding (metal not specified) (Ref. 19), and the other paper pointing out a species specificity of response to welding gases (Ref. 235).

Acute

In experiments by Von Haam and Groom, 1941 (Ref. 235), 3 mice (sex, strain, and age not provided) were exposed for 6 hours to 1600 to 2600 mg/m^3 of welding fumes from A5 electrodes (17.6% Fe_2O_3, 9.5% Cr_2O_3, 10.6% Mn_3O_4, 16.8% CaO, 16.1% Na_2O, 5% Al_2O_3, 14.6% F). The mice showed no abnormal lung reaction. All 4 mice exposed to filtered welding fumes (particulates removed; 19.5% O_2, 78% N_2, 0.7% CO_2, trace CO, and qualitative confirmation of presence of nitrogen oxides) died; the exposure duration was not provided. Mice were stated by the authors to possess the least resistance to welding fumes and gases of all the species they tested (rats, rabbits, and guinea pigs).

Chronic

Erman and Rappoport, 1970 (Ref. 19), exposed albino mice (number of mice, sex, and age not specified) to welding fumes for 4 hours daily for 6 months. Fumes originated from arc welding using CO_2 as a shielding gas and contained 150 to 180 mg/m^3 dust, 122 to 140 ppm CO, 3.6 to 6.8 mg/m^3 nitrogen oxides (1.92 to 3.62 ppm based on NO_2), and 0.18 ppm ozone. Decreased body weight gain, increase in relative kidney and liver weight, central nervous system effects, depression of cholinesterase activity, degeneration of nerve cells in the cerebral cortex and subcortical region, moderate myocardial cell degeneration and thickening of the stromal tissues of the heart, liver, and kidneys, and slight pulmonary fibrosis were noted. The general toxicity of the fumes was greater than the degree of lung changes noted.

Effects in Cats

Titus et al., 1935 (Ref. 246), studied the acute pulmonary effects of exposing cats to fumes and gases from dc arc cutting of iron. Bare iron electrodes were used to cut "blue annealed steel" (SAE 10-15), generating welding fumes that contained ferric oxide as the main component. Ferric oxide fume concentrations ranged from 60 to 250 mg/m^3. Animals were exposed for 175 to 369 minutes. Carbon dioxide (1 to 14 percent) was added to the air in the exposure chamber to increase the respiration rate of the cats.

A total of 11 cats (sex, age, and weight not given) were used in 5 experiments. Table D11 presents the exposure and results of these experiments. The cats were observed to develop varying degrees of pulmonary edema and inflammation.

Intratracheal Administration of Welding Fume Suspensions

The following studies, all performed in rats, illustrate the ability of suspensions of welding dusts to provoke phagocytosis (Ref. 20) and fibrosis, in the case of dust from acid covered electrodes (Ref. 25). Rutile covered electrode dusts caused increases in collagen and ascorbic acid content of rat lungs (Ref. 24). Increases in both serum neuraminic acid levels and certain amino acids occurred in rats that were intratracheally administered suspensions of welding fumes from low fluoride covered electrodes (Refs. 249 and 250). Details of these experiments are presented below and are summarized in Table 38 in Chapter 3 of the main text.

The effects of welding dusts from rutile, basic, and acidic covered electrodes on rat lung levels of hydroxyproline, total lipids, and phospholipids, which are indicators of fibrogenesis, were reported by Kysela et al., 1973 (Ref. 25). The dust (50 mg/ml saline; particle size under 5 μ) was administered intratracheally to 15 male Wistar rats (180 to 200 g; age not given). The animals were killed 12 or 24 weeks after treatment and their

Table D11
Effects of exposure of cats to welding fumes

Number of cats	Av. conc. of Fe_2O_3 (mg/m^3)	Exposure time (min.)	CO_2 conc. (%)	Results
3	99	227	"high"	2 died; lungs edematous and red or pigmented
3	155	175	none	1 normal; 1 slight pulmonary edema; 1 used in next experiment
4	60	207	<10	2 died with pulmonary edema; 1 with severe edema at autopsy; 1 survived and used in next experiment
1	250	369	1-7	Died of pulmonary edema, darkly pigmented lungs, inflamed trachea
2	250	312	none	1 edematous, 1 normal 8 days later (autopsied)

Titus et al., 1935 (Ref. 246).

lungs analyzed for hydroxyproline, total lipids, and phospholipids.

Welding fumes from basic covered electrodes were highly toxic; all exposed animals died within 24 hours. Autopsy showed congestion and reduced aeration of the lung tissue, which the authors felt might have been caused by the high fluoride content of the electrode.

Administration of various dusts into the lungs of rats increased the wet weight of the lungs and the levels of hydroxyproline, total lipids, and phospholipids in the lung tissue. Acid covered electrode dusts were of greater fibrogenicity than dusts from rutile covered electrodes. Histological examination showed nodule formation in lung tissues of exposed animals, with collagen fibers on the periphery of some nodules.

Intratracheal insufflation of welding fumes generated by burning covered electrodes low in fluorine content (EP-50-BNT) was found to increase proline and hydroxyproline levels in the lungs of rats according to studies by Senczuk and Nater, 1970 (Ref. 26). The electrode coating consisted of 6% CaF_2, 5% SiO_2, and 2% Mn. Male Wistar rats (10 to 12 per group, about 2 months old, 150 to 160 g) were intratracheally administered 20 mg of welding dust in 1 ml of saline. A control group received 1 ml saline only. The levels of hydroxyproline and proline in the lung tissues were determined 5, 10, 21, 42, and 84 days after treatment. Lungs of the exposed rats contained significantly higher levels than controls of both hydroxyproline (4.7 mg after 84 days; controls, 3.2 mg) and proline (8.6 mg after 42 days; controls, 4.1 mg). Other changes noted in experimental rats included increased lung weights and reduced rates of weight gain, as shown in Table D12.

The effect of welding fumes on the lung content of collagen and ascorbic acid, indices of lung damage and repair in rats, was reported by Naumenko, 1966 (Ref. 24). Welding fumes generated from ANO-1 rutile covered electrodes (composition not specified) were administered intratracheally in single doses (50 mg in 1 ml saline) to white rats (strain, sex, age, and number not given). The fumes consisted of oxides of manganese, iron, silicon, and other components (percentages not given); 85 percent of the particles were 0.5 μ in diameter or less. Control rats were administered 1 ml saline only. The animals were killed 7, 30, 90, 180, and 270 days after administration of the dust and their lungs were analyzed for collagen and ascorbic acid. Results are shown in Table D13.

There was a progressive rise in collagen with respect to lung weight; the relative content of collagen as well as ascorbic acid in the lung tissue reached a maximum by the sixth month of exposure.

The effects of inhalation of welding fumes on the levels of amino acids and glycoproteins in blood serum of rats were studied by Senczuk et al., 1970 (Refs. 249 and 250). Male white Wistar rats (150 to 180 g) were treated by intratracheal insufflation with welding fumes obtained from burning covered electrodes of low fluoride content (Polish EP-50-BNT and EP-49-20). Fumes contained unspecified concentrations of iron, manganese, titanium, fluorides, silicon, and zinc. The first group of animals (75 rats for the amino acid experiment and 80 rats for the glycoprotein study) received 20 mg of fumes in 1 ml saline intratracheally; the second group (50 rats and 70 rats for amino acids and glycoprotein studies, respectively) received 1 ml saline intratracheally; the third group (12 and 20 rats) remained untreated. The rats were killed after 5, 10, 21, 42, and 84 days, and the levels of amino acids (glycine, methionine, serine, cysteine, cystine, and taurine) and glycoproteins

Table D12
Body weight gain and lung weights of Wistar rats following administration of welding fume suspension

Days after exposure	5	10	21	42	84	168
Percent body wt. gain in controls	0.99	7.8	4.3	39.0	60.0	---
Percent body wt. gain in exposed rats	1.8	1.9	2.0	23.0	35.0	---
Avg. wt. increase in lungs of controls (mg)	---	0.23	0.23	0.31	0.26	0.30
Avg. wt. increase in lungs of exposed rats	---	0.24	0.28	0.33	0.33	0.41

Senczuk and Nater, 1970 (Ref. 26).

Table D13
Lung collagen and ascorbic acid content of rats following intratracheal administration of welding fume suspension

Days post exposure	7	30	90	180	270	Control
Collagen (mg/lung)	6.2	7.2	18.8	20.9	36.4	9.1-15.9
Collagen (mg/100 mg tissue)	1.1	1.5	2.7	4.4	4.4	2.7-3.0
Ascorbic acid (mg%)	26.0	29.3	32.9	33.8	28.9	20.3-22.2

Naumenko, 1966 (Ref. 24).

(glucosamine, seromucoid, and neuraminic acid) in blood serum were determined.

By 21 days after treatment, the levels of glycine, serine, methionine, cystine, cysteine, and taurine increased in blood serum. After 42 and 84 days, the levels of glycine, methionine, and serine decreased below the control values; only traces of cystine and taurine, but not cysteine, were detected (Ref. 249).

There was an initial elevation in the content of seromucoid and neuraminic acid in blood serum 5 days after treatment. The level of seromucoid decreased below that of control values 21 days later. Neuraminic acid levels dropped rapidly 10 days after treatment and approached the control values 42 days later. There was no significant change in the content of glucosamine in serum after the intratracheal insufflation of welding fumes (Ref. 250).

The fibrogenicity and ease of phagocytosis of welding dusts from two types of rutile covered electrodes (OZS-4 and TsL-11, composition not given) in the lungs of rats were reported by Guskova and Komovnikov, 1974 (Ref. 20). The fume concentrate (290 to 310 mg/m^3; 95 percent of particles $< 5\ \mu$ in diameter) was administered intratracheally in a single dose of 50 mg to 100 white rats (strain, sex, age, and number per group not given). Animals were killed and examined after intervals of 3, 6, 9, and 12 months.

Results showed that both the number of active alveolar phagocytes and phagocytic index were increased shortly after treatment and the values gradually returned to the control level after 12 months. The effects were more pronounced for dust from electrode OZS-4 than that from TsL-11. The number of destroyed cells that contained dust was also high shortly after treatment for dust from electrode OZS-4 and returned to normal 12 months later. Intratracheally administered dust from TsL-11 was associated with increased numbers of destroyed dust cells 3 to 6 months after treatment.

Fibrotic changes were evaluated by determining collagen content (method not specified); the lungs of rats exposed to welding fumes had increasingly higher collagen contents than those of the controls, as shown in Table D14.

The authors concluded that these welding dusts were phagocytized well by the lungs and were of relatively low fibrogenicity.

Table D14
Lung collagen content of rats intratracheally administered welding fume suspension

Months post admin.	Lung collagen (mg)		
	Control	Intratracheal OZS-4	Intratracheal TsL-11
3	9.1	18.3	16.1
6	9.7	20.4	16.8
9	10.4	21.5	18.0
12	12.4	25.3	19.6

Guskova and Komovnikov, 1974 (Ref. 20).

Distribution and Excretion of Inhaled Welding Fumes

This section presents results of inhalation experiments in various laboratory animals in which the distribution and excretion of inhaled welding particles were examined.

Rats and rabbits inhaling welding fumes containing 17 percent calcium fluoride from basic covered EP 49-29 Polish electrodes accumulated fluorides in the bones and teeth (from initial lung uptake) and excreted fluorides in urine and feces, resulting in constant blood fluoride levels (Ref. 232). Elevated manganese and silica levels were noted in the lungs, liver, and kidneys of guinea pigs exposed to welding fumes from EP 52-28P and EP 47-28P Polish electrodes (Refs. 252 through 254). Temporal excretion patterns for manganese, magnesium, and iron were studied in rats exposed to mild steel welding fumes. The lungs were observed to retain manganese (Ref. 302). Iron, cobalt, chromium, and antimony levels in tissues of rats inhaling rutile covered electrode welding fumes were also investigated. Chromium was eliminated from the lungs more slowly than the other metals (Refs. 237 and 238). It was found that in the mouse, excretion of titanium, chromium, manganese, iron, and nickel inhaled in welding fumes from stainless steel electrodes required more than 24 hours (Ref. 303).

Details of these investigations follow.

Fluorides

Krechniak, 1969 (Ref. 232), investigated the fluoride content of animal tissues after long-term exposure to welding fumes from basic covered electrodes containing 17 percent calcium fluoride (Polish electrode EP 49-29). The average fluoride content of the fume was 8.9 percent.

Rats were exposed to these fumes for 95 days, rabbits for 110 days, three hours daily; 150 white rats (150 to 220 g) and 45 one-year-old rabbits (3 to 5 kg) were used. Average fluoride content of the fumes was 60.6 mg/m^3, particulate fluorides averaged 3.4 mg/m^3, and volatile fluorine compounds 1.45 mg/m^3 (1.75 ppm) calculated as HF. One electrode was burned every 10 minutes. Animals were killed approximately every other week during the exposure; some were allowed to recover for 130 days after the total exposure and were then killed. Fluoride contents of the blood, muscles, liver, kidney, brain, lung, trachea, femur, and teeth were analyzed by the method of Fabre. The results of tissue analyses in rats and rabbits are presented in Tables D15 and D16, respectively. The fluoride content of the lung increased rapidly over the first 3 days of exposure and remained elevated even after 20 weeks of recovery after the end of exposure. Fluoride levels were significantly higher than in the controls in the teeth after 10 days in rats and 14 days in rabbits, and also in bones after

Table D15
Fluoride content of tissues of rats exposed to welding fumes

Exposure (days)	F content (ppm)								
	Blood	Muscle	Liver	Kidney	Brain	Lung	Trachea	Femur	Mandible
0 (Control)	0.6	0.5	0.5	2.2	1.7	1.1	5.5	162	176
3	---	0.6	---	---	---	3.4*	6.6	---	---
10	---	1.0	---	---	---	3.1*	6.3	---	202*
20	---	---	---	---	---	3.1*	18.5*	---	373*
30	1.5*	1.1*	2.2*	3.2*	5.0*	3.4*	---	350*	382*
65	---	2.6*	2.4*	5.7*	7.0*	4.7*	24.0*	586*	709*
90	1.7*	2.3*	2.6*	6.0*	7.2*	5.7*	27.9*	791*	1030*
130 (Recovery)	0.9*	1.0	0.9	2.5	1.8	3.2*	13.1*	687*	793*

*Statistically significant elevation ($p < 0.05$).

Krechniak, 1969 (Ref. 232).

Table D16
Fluoride content of tissues of rabbits exposed to welding fumes

Exposure (days)	Blood	Muscle	Liver	Kidney	F content (ppm) Brain	Lung	Trachea	Femur	Teeth
0 (Control)	0.7	1.1	0.5	0.8	0.8	1.2	1.8	130	182
3	1.5*	---	---	---	---	2.7*	---	132	182
14	1.0*	1.1	1.2	1.7	1.4	2.2*	2.0	162	204*
28	1.2*	1.8*	1.9*	1.9*	1.8*	3.9*	9.6*	146	234*
40	1.9*	1.1	0.6	1.9	1.4*	2.1*	11.7*	253*	328*
56	1.0*	2.7*	1.8*	1.6*	2.3*	2.9*	16.1*	411*	530*
110	1.2*	1.7*	1.2*	1.7*	1.6*	3.2*	17.6*	569*	676*
140 (Recovery)	1.0*	1.1	0.6	1.2	1.1	2.3*	9.0*	269*	362*

*Statistically significant elevation ($p < 0.05$).

Krechniak, 1969 (Ref. 232).

30 days of exposure of rats and 40 days in rabbits, remaining elevated after the end of exposure. Fluorides were said to have been distributed from their initial uptake in lungs through the bloodstream, where they were removed to bones and teeth or excreted in urine and feces, thus keeping the blood level constant during the exposure period. Urinary and fecal fluoride levels were also monitored in the rabbits. After 14 days of exposure, and then continuing to the end of the experiment, a urinary fluoride level several times higher than normal was maintained, as shown in Table D17.

Table D17
Urinary and fecal fluoride levels in rabbits exposed to welding fumes

Exposure (days)	Mean urinary fluorides (ppm)	Mean fecal fluorides (ppm)
0 (control)	1.3	13.9
3	2.0*	22
7	2.9*	40*
14	5.2*	40*
28	6.2*	50*
40	4.2*	30*
56	4.0*	32*
90 (recovery)	5.4*	34*

* = statistically significant elevation ($p < 0.05$)

Krechniak, 1969 (Ref. 232).

Manganese

Kukula et al., 1968 (Ref. 252), exposed guinea pigs to welding fumes from EP 47-28P electrodes manufactured in Poland (composition of electrode not specified), in order to study changes on manganese levels of the lungs, liver, kidneys, brain, and pancreas of the exposed animals. Twenty-five animals (sex, strain, and age not provided) were exposed to welding fumes containing 2.52 mg/m^3 manganese oxides, 4 hours daily, for 126 days. Another 25 guinea pigs were exposed for the same duration to 1.72 mg/m^3 of manganese oxides from welding fumes. There were 25 unexposed controls. Some animals were killed after the exposure, while others were allowed a 134-day recovery period before they were killed. Tissues were analyzed for manganese content, measured as μg/g dry tissue weight; results are presented in Table D18.

The authors noted that lung manganese levels, but not liver or kidney levels, depended upon the duration of exposure as well as the fume level of manganese. In the guinea pigs inhaling the higher manganese fume level, lung manganese content progressively rose from 0.38 μg/g dry tissue before exposure to 70.6 μg/g in animals killed after 7 days of exposure, 254 μg/g in animals killed after 53 days, and up to 494.1 μg/g in animals killed after the full 126 days of exposure.

Kolanecki and Skucinski, 1968 (Ref. 253), measured the manganese content of brain and pancreas tissue from guinea pigs exposed to welding fumes from two Polish manufactured electrodes stated by the authors to be of similar manganese content (level not specified): EP 47-28P and EP 52-28P electrodes. Two manganese fume levels were used in the exposure: 1.44 mg/m^3 and 2.37 mg/m^3. It was unclear which electrode type was used to achieve which fume level, however. Guinea pigs

Table D18
Lung, liver, and kidney manganese levels in guinea pigs exposed to welding fumes

	Lungs	Liver (μg/g dry tissue)	Kidneys
Control	0.5	4.1	1.9
126 days of exposure to 1.72 mg/m^3 Mn	370.0	6.2	4.0
126 days of exposure to 2.52 mg/m^3 Mn	494.1	7.5	4.4
134 days recovery following 126 day exposure to 1.72 mg/m^3 Mn	171.6	5.1	2.4
134 days recovery following 126 day exposure to 2.52 mg/m^3 Mn	145.1	4.5	2.3

Kukula et al., 1968 (Ref. 252).

(strain, sex, age, and number of animals not provided) were exposed to either fume level for 4 hours daily for up to 6 months and killed after the exposure, while others were exposed for 126 days and allowed to recover for 136 days prior to being killed. Brain and pancreas manganese contents, in μg/g of dry tissue, are presented in Table D19. Both brain and pancreatic manganese levels were elevated by 126 days of exposure. Only brain levels returned to normal following the 136 day recovery period; pancreatic manganese levels remained elevated.

Silica

Bunka and Kiziewiez, 1968 (Ref. 254), found that the SiO_2 content in the lungs, liver, and kidney increased rapidly in 50 guinea pigs (sex, strain, and age not specified) exposed to welding fumes from either of two similar Polish electrodes, EP 47-28P or EP 52-28P (4.1 to 4.3 mg/m^3 SiO_2), for up to 126 days, 4 hours daily, and killed after exposure. Some animals (number not specified) were allowed to recover after exposure for 40 or 136 days prior to being killed. There were 25 controls. The high SiO_2 content in exposed guinea pigs decreased rapidly in the lungs, but less rapidly in liver and kidneys, as shown in Table D20.

Mixed Fume Studies

Leslie et al., 1976 (Ref. 303), exposed 4 adult white mice for 4 hours to fumes from arc welding a 2 cm mild steel base plate with Type 308-16 stainless steel electrodes. Ten rods per hour were consumed. Fume particle size averaged 0.5 to 1.0 μ in diameter; the composition of the welding fume particles in this size range included 16.9 μg/m^3 titanium, 101.0 μg/m^3 chromium, 107.0 μg/m^3 manganese, 102.0 μg/m^3 iron, and 8.4 μg/m^3 of nickel. Humidity was maintained between 70 and 80 percent.

The elements present in the fume could be clearly detected in the lung tissue of the exposed mice using proton-induced x-ray emission (PIXE) for elemental analysis. No evidence of elevated levels of Fe, Ti, Cr, or Mn were found in blood samples taken immediately after exposure, or 1, 4, or 24 hours later. The levels of Ti, Cr, Mn, and Ni remained consistent within each mouse lung over a 24-hour period, indicating that a time frame longer than 24 hours is required before significant excretion of the uptake occurs.

Hewitt and Hicks, 1973 (Ref. 237), and Hewitt and Hicks, 1972 (Ref. 238), investigated the effects of inhaled welding fumes and the retention of metallic elements in tissues of rats. Arc welding fumes were produced by igniting a rutile covered iron welding rod (Phillip Type C18) on a workpiece of medium tensile strength steel. Copper was present in the welding rod, and the workpiece contained chromium, antimony, cobalt, lead, and a small amount of zinc. The electrode coating consisted of limestone, manganese dioxide, kaoline and cellulose powder bound by sodium, and potassium silicates. A new welding rod was burned every 3 minutes to maintain the fume concentration at an average of 1500 mg/m^3. The welding fumes, which comprised decomposition

Table D19
Brain and pancreas manganese levels in guinea pigs exposed to welding fumes

	Brain	(μg/g dry tissue)	Pancreas
Control	3.25		4.28
3 months of exposure to 1.44 mg/m^3 Mn	7.34		8.42
3 months of exposure to 2.37 mg/m^3 Mn	5.32		8.42
6 months of exposure to 1.44 mg/m^3 Mn	4.53		---
6 months of exposure to 2.37 mg/m^3 Man	4.87		---
126 days of exposure to 1.44 mg/m^3 Mn plus 136 days of recovery	3.87		12.54
126 days of exposure to 2.37 mg/m^3 Mn plus 136 days of recovery	3.36		10.91

Kolanecki and Skucinski, 1968 (Ref. 253).

Table D20
Lung, liver, and kidney content of SiO_2 in guinea pigs exposed to welding fumes

		Lung	Liver (μg/g of dry tissue)	Kidney
Control		76-90	35-41	27-33
Exposure	7 days	251	45	27
	12 days	259	33	37
	53 days	663	69	47
	90 days	840	106	68
	126 days	996	146	85
Recovery	40 days	425	97	65
	134 days	135	28	38

Bunka and Kiziewiez, 1968 (Ref. 254)

products of both the welding rod and the workpiece, were drawn into a specially constructed exposure chamber at an air flow rate of 0.35 m^3/min. The air temperature in the chamber did not exceed 25° C at any time.

Male albino rats, CSE strain (about 10 weeks old, 200 to 250 g, free from detectable respiratory infection) were used for these investigations. In the first experiment, 9 rats were placed in the exposure chamber; 2 animals were removed after 30 minutes of exposure, the rest after 4 hours. Nine rats from the same colony were used as controls. Twenty-four hours after the completion of the exposure, all the animals were weighed, examined externally, and then killed. Blood samples were taken immediately by cardiac puncture, and tissues of organs were prepared for histological examination. Well-cleaned polythene equipment was used to prevent metal contamination of the specimens. In the second experiment, 8 rats were placed in the exposure chamber for 4 hours, and 2 rats were killed at 1, 7, 28, and 75 days after the exposure. Pairs of control rats were killed at the same time as the exposed animals. Autopsy examinations were performed, and lung tissues were prepared for histological examination.

Neutron activation analysis, according to methods of Hewitt and Hicks, 1977 (Ref. 304), was used to determine the concentration of several elements in a single small sample without chemical separation. Absorption spectrophotometry was used for lead, manganese, and copper determinations because lead does not form a convenient radionuclide on neutron irradiation, and the half lives of radioactive ^{56}Mn and ^{64}Cu are too short for accurate measurement. The silica content of the fumes cannot be determined by neutron activation analysis and was estimated gravimetrically after removal of iron and organic material. The concentrations (mg/m^3) of some of the elements in the welding fume were: iron, 500; silica (as SiO_2), 380; manganese, 95; lead, 7.0; copper, 1.0; antimony, 1.0; cobalt, 0.4; chromium, 0.1; and zinc, < 0.1. The mean particle size of the fume was found to be 0.15 μm, as determined by electron microscopy.

Results of the 30 min to 4 hour exposure included significant ($p < 0.05$) increases in the levels of iron, cobalt, chromium, and antimony in the lung tissues of exposed rats as compared to the controls. The increase in concentration in the lungs was more pronounced after 4 hours than after 30 minutes of exposure to the welding fumes. The levels of cobalt increased about fourfold in the blood and twofold in liver tissue in the exposed animals as compared to those of the control rats. No significant changes in the concentrations of iron, chromium, or antimony were detected in the blood or liver of these animals. In kidney tissues, the levels of iron, cobalt, chromium, or antimony from exposed rats were not significantly different from those of the controls.

Rats exposed for 4 hours, then killed 1, 7, 28, and 75 days later, showed that metallic elements detected in the lung tissue were being eliminated at different rates. The concentrations of antimony and cobalt in the lungs decreased rapidly during the first week after exposure. The chromium concentration was not very high in the lungs of rats 1 day after the exposure to welding fumes, and its elimination was less pronounced compared to that of antimony or cobalt. The disappearance of iron proceeded in a biphasic manner: an initial rapid elimination (half life about 1 day) over the first few days, followed by a slow removal (half life about 33 days). Tissue levels of silica, manganese, or lead were not reported.

Ishimi and Ohmoto, 1974 (Ref. 302), studied the distribution of iron, manganese, and magnesium in rats inhaling welding fumes. The welding fume was generated from SS-41 mild steel with a 4 mm electrode of an ac welding machine, using a current of 165 A, and a welding rate of 20 cm/min. A fume density of 500 mg/m^3 with 17.1% Fe, 4.5% Mn, and 0.06% Mg was noted.

In unexposed Wistar rats, iron content measured by atomic absorption spectrophotometry was highest in the lungs, liver, kidney, bladder, brain, stomach, small intestine, and muscle tissue. Manganese content was high in the liver and very low in the blood. Magnesium content was relatively greater in muscle than in the liver, and low in other tissues.

Twenty Wistar rats were exposed to these welding fumes for 40 minutes. Ten that were killed 1 to 3 hours after exposure showed increased iron content in the lung, stomach, small intestine, blood, and feces. In 5 animals killed 24 hours after exposure, there was a decrease in iron content in these tissues and feces, and for the 5 killed 6 days after exposure, iron levels were higher than controls only for lungs and feces. Iron levels in brain, liver, and muscle did not show these temporal changes. Temporal changes in manganese content were similar to those for iron at 3 hours after inhalation. Bladder and fecal manganese levels were highest at 24 hours. Lung manganese content had not returned to control levels one week after exposure. The magnesium distribution was similar to that of iron and manganese.

Inhalation of 500 mg/m^3 of the welding fume for 40 minutes allowed distribution of fume constituents to various organs, which approached peak levels one hour following exposure. Absorption from the lungs continued thereafter but was controlled by excretion of the fume component. Iron uptake from the lungs into the bloodstream was greatest 3 hours after exposure. In the case of manganese, the lung content was elevated even 24 hours after the inhalation of welding fume, and recovery from the metal retention was not observed even 1 week after the blood concentration of manganese returned to control levels, thus indicating that the lungs did not release inhaled manganese. The brain showed almost no uptake of iron, manganese, or magnesium. There was no retention of these metals in liver or kidneys over the one-week period (Ref. 302).

Carcinogenicity Studies

Inhalation and intratracheal administration of dust generated from burning electrodes or welding has caused edema and varying degrees of pulmonary damage in experimental animals, as shown in the previous section. However, lung tumors have not been reported as an incidental finding following this treatment. One investigation, which sought for lung tumors in rats exposed to dust from chromium-containing electrodes, failed to find tumors up to 1.5 years after exposure (Ref. 234). No other research reports of tumors following exposure to welding fumes, gases, or dusts were available. Further details of the cancer study are presented as follows.

Magai and Norkin, 1965 (Ref. 234), intratracheally administered a suspension of 50 mg of dust from welding with 606/11 electrodes to each of 10 rats. The atmosphere from which the dust was collected contained 20.2 mg/m^3 MnO_2, 22.5 mg/m^3 CrO_3, 32.1 mg/m^3 F, and 4 mg/m^3

SiF_4. These rats, killed 1.5 years after the treatment, showed no evidence of lung tumors. Another 10 rats that were exposed to welding fumes for 9 months similarly revealed no formation of lung tumors in 1.5 years. The schedule of exposures, sex, strain, and age of the animals were not given.

Intraperitoneal and Subcutaneous Administration of Welding Fume Suspensions

This section presents the results of experiments in which welding fume suspensions, solutions, or both were injected either intraperitoneally or subcutaneously into various species of experimental animals.

Intraperitoneal injections of suspensions of welding dust have been found to stimulate phagocytosis in mice and rats (Refs. 234 and 240). Effects in guinea pigs depended upon the type of electrode used to produce the dust as well as the solubility of the dust particulates. Suspensions of water soluble dusts from basic covered LB-53 electrodes caused complete mortality within one hour of intraperitoneal injection in guinea pigs due to intraperitoneal bleeding. Injections of insoluble dust suspensions did not cause death, however (Ref. 247).

Blood vessel permeability alterations were noted in rabbits following subcutaneous injections of solutions of water soluble welding dust from basic covered LB-52 electrodes and ilmenite covered B-17 electrodes (Ref. 247).

Details of these investigations are presented as follows.

Intraperitoneal Effects

Leonicheva, 1965 (Ref. 240), intraperitoneally injected 200 white mice (strain and age not specified) with an unspecified dose of a suspension of dusts from welding either aluminum or a magnesium alloy of aluminum. A group of control mice was injected with alumina powder and the phagocytic response was studied. Neutrophil phagocytosis was most active 3 hours after administration and subsided after 5 days, while macrophage phagocytosis had maximum activity 3 days after injection. Alumina powder was readily phagocytosed; no dust particles were found in the peritoneal fluid by the 5th day. However, Al-Mg welding dust was least readily phagocytosed. Therefore, Al welding fume particles were stated to be more pathogenic than dust from Al-Mg welding.

Migai and Norkin, 1965 (Ref. 234), studied the effects of welding substances on phagocytosis, using 20 rats (strain, sex, and age not given) and injecting them intraperitoneally with suspensions of 25 mg/ml saline of welding fumes generated by electrodes 606/11 and UONI-13/45. The electrode fume compositions are listed in Table D1. Twenty control rats received coal dust injections. At intervals of 5, 24, 48, 72, and 96 hours, peritoneal fluid samples were taken and examined for phagocytosis. Welding dust from electrode 606/11 caused a higher phagocytosis index than electrode UONI-13/45; both were higher than that of the controls.

The inhalation experiments reported by Kawada et al., 1964 (Ref. 247), revealed that the particulate portion of the welding fumes, not the gases, was responsible for the toxic effects on guinea pigs (see the section on inhalation of welding fumes). The welding dust suspension, as well as the water soluble and insoluble components of it, were therefore tested by intraperitoneal injection in guinea pigs. Samples were prepared by suspending 0.15 g of the welding dust from burning three types of electrodes (basic covered LB-52, basic covered LBM-52, and ilmenite covered B-17; composition specified in Table D9) in 2 ml distilled water. For the water soluble and insoluble components, the suspension was centrifuged at 3000 rpm for 10 minutes, and the supernatant and sediment were separated and made up to 2 ml. The samples were sterilized and injected into the peritoneal cavities of the guinea pigs.

All 9 guinea pigs injected with welding dust suspension from the basic covered electrode LB-53 died within 3.5 hours, the shortest survival time being 15 minutes, whereas none of the 6 guinea pigs injected with dust suspension from the ilmenite covered electrode B-17 died. Only 3 out of 12 receiving dust suspension from the improved basic covered electrode LBM-52 died.

The water soluble component of the welding dust from electrode LB-53 was found to be very toxic; all 6 guinea pigs receiving such injections died within one hour, the shortest survival time being 17 minutes, whereas all animals (number not given) injected with the water soluble sediment suspension survived. All the dead animals showed bleeding into the peritoneal cavity.

Repeating the intraperitoneal injection experiments with a total of 14 types of electrodes with various compositions, it was concluded that fluorine and potassium (KF, KHF_2, KOH, and NaF) in the basic electrode coatings were the principle toxic substances, causing death, severe toxicity, or both (no experimental details were provided).

Subcutaneous Effects

Kawada et al., 1964 (Ref. 247), tested the effects of the water soluble components of welding particulates from LB-52 basic covered electrodes and B-17 ilmenite covered electrodes on the blood vessel permeability of rabbits. (Tables D9 and D10 specify electrode and fume compositions.) The solution was prepared by centrifuging a suspension of 0.15 g welding dust in 2 ml water, followed by sterilization by boiling. The supernatant solution (quantity not given) was injected subcutaneously into the abdominal region of groups of 5 rabbits (weight and strain not given) followed immediately by an intravenous injection of a 1 percent solution of tripan blue (volume not given). The time required for the first appearance of bluish tint at the injection site and the degree of coloration were compared to determine changes in blood vessel permeability. The appearance of the dye in rabbits injected with water soluble components of the welding dust from the basic covered electrode LB-52 was faster (average 5 minutes) than in those injected with a similar sample from the ilmenite covered B-17 electrode (average 80 minutes).

Various chemical compounds of coated electrodes were also tested for their effects on the permeability of blood vessels. The shortest time for the appearance of the injected dye was produced by the NaOH-containing electrodes (average 12.5 minutes), followed by MgO, K_2CO_3, NaF, KOH, KF, CaF_2, KHF_2, CaO, $NaHF_2$, SiO_2, Na_2SiF_6, and K_2SiF_6. The intensity of the dye was higher in animals injected with MgO, NaF, KF, CaO, and KHF_2 than those receiving NaOH, K_2CO_3, or KOH (details of experiments and results were not presented).

Metal Fume Fever

Almost no animal research on causal mechanisms and dynamics of metal fume fever has been reported, although most authors working with human metal fume fever cases admit that research is sorely needed. The cause of the fever seems to lie in the realm of an immunologic response to metal complexes formed in the lung. The cause of an often noted temporary immunity to the flu-like disease is much farther from being elucidated. These topics are more fully covered in Chapter 2 of this report.

As far as animal experimentation is concerned, one paper is reviewed in this section. The authors attempted to produce metal fume fever in rabbits, monitoring body temperature and white blood cell counts after exposure, as detailed in the following discussion.

Kawada et al., 1964 (Ref. 247), investigated the ability of welding fumes to cause body temperature elevation in mature male rabbits, similar to metal fume fever in humans. They took the rectal temperature before exposure and after 0.5 to 1.0 hour of inhalation; they recorded the temperature of the rabbits (strain not provided) continuously for 10 hours or more with a thermistor. Three control groups were used: 5 rabbits were not exposed to any welding fumes; 10 rabbits were exposed to zinc fumes produced by an arc between zinc powder on a steel plate and gouging carbon, as a positive control; and 2 rabbits inhaled gases and fumes from an arc between the steel plate and gouging carbon without zinc powder.

The average body temperature change in 6 hours for the 5 control rabbits was −0.2° C, and for the 2 rabbits inhaling arc gas without zinc powder were from +0.5° C to −0.2° C. The temperature of 8 of the 10 rabbits fell (increment not given) immediately after inhaling zinc fumes, then rose slowly, reaching maximum value 6 to 8 hours later (average +0.51° C, maximum +1.3° C) and falling again thereafter. For the 5 rabbits exposed to welding fumes and gases from ilmenite covered B-17 electrodes and the six exposed to fumes and gases from basic covered LBM-52 electrodes, the average changes in body temperature were +0.04° C and +0.13° C, respectively. Immediately after exposure, the temperature of the 6 rabbits inhaling welding fumes and gases from the basic covered LB-52 electrode fell and then rose gradually 3 to 4 hours later. The average rise in temperature was +0.38° C for this group.

Leukocyte counts increased over 10 percent in 3 of 6 animals exposed to fumes from the LB-52 electrodes; increased white blood cell counts also were noted in one each of rabbits exposed to the B-17 and LBM-52 electrode fumes and gases.

The authors emphasized the weak effect of smoke from LBM-52 basic covered electrodes in causing fever and leukocytosis in rabbits (similar to metal fume fever in humans). They noted that fumes and gases from welding with LB-52 basic covered electrodes has a slightly stronger pyrogenic effect in rabbits. However, no statistical comparison of the control and treated rabbits' temperatures was presented to elucidate the significance of these slight temperature changes.

Effects of Welding Fumes on Influenza or Pulmonary Tuberculosis

In this section, two articles are presented. The first deals with a possible association between intratracheally administered welding fume suspensions and influenza virus pathogenicity in the lungs of rats (Ref. 255). The second reveals no synergistic activity between inhaled welding fumes and pulmonary tuberculosis in guinea pigs, rats, or mice (Ref. 258). These studies were probably undertaken in response to epidemiologic indications at the time when welders died more frequently of lung infections than nonwelders; these data are presented in Chapter 2 of the report.

Influenza

The combined effects of welding fumes and influenza virus on lung tissue of rats were studied by Naumenko and Frolov, 1968 (Ref. 255). Three groups of white rats (180 to 200 g) were used for the experiment. The first group (58 rats) received one intratracheal injection of 50 mg of welding fumes (particle size 0.5 to 2 μ) in 1 ml of saline. The welding fumes consisted of oxides of iron, manganese, silica, and other unspecified components. A second group of 56 rats received 0.2 ml of influenza virus suspension (strains A, A_2, B_{14}, and PP_8; biological titer 10^{-5}) intratracheally at 10-day intervals. The third group (66 rats) received both welding fumes and influenza virus suspension. All animals were kept under observation and the nucleic acid content of lung tissues was analyzed after 40, 100, 180, and 270 days.

Results showed no relationship between the quantities of DNA or RNA in lung tissue and the length of the observation period. However, the content of DNA and RNA in the lung and the ratio of RNA to DNA were generally highest in the group of rats receiving a combination of welding fumes and influenza virus and lowest in the group receiving the virus alone, shown as follows.

Treatment	DNA (μg)	RNA (μg)	RNA/DNA
Welding fumes	254-325	108-150	0.34-0.55
Influenza virus	262-294	86-93	0.29-0.35
Welding fumes plus influenza virus	294-314	137-216	0.44-0.73

The authors suggested a possible association of welding fumes and influenza virus in the pathogenesis of welding fume toxicity.

Tuberculosis

The synergistic pulmonary effects of welding fumes and tuberculosis (TB) infection in guinea pigs, rats, and mice were studied by Gardner and McCrum, 1942 (Ref. 256). Repeated exposure to welding fumes did not increase susceptibility to TB infection, nor did it reactivate partially healed pulmonary tubercles to produce progressive TB.

Welding fumes were generated by burning heavily coated mild steel electrode rods, 5/32 in. diameter. Fumes contained high levels of Fe_3O_4, low levels of Mn and Al, and SiO_2 (none as quartz). No fluorides were present in the coating. Fume concentrations in the animal exposure chamber were maintained by burning 12 to 33 electrodes per day.

Five groups of animal experiments were performed: Group I–Animals were exposed only to welding fumes for up to 12 months (50 guinea pigs, 300 to 600 g; 25 white rats and 12 white mice, age, sex, and weight not given). Group II–Animals were infected with TB only (50 guinea pigs). Group III–Animals were exposed to welding fumes for 3 days, then infected with TB (50 guinea pigs). Group IV–Animals were infected with TB first; then, during the healing period of 4, 5, 6, 8, or 10 months after infection, these animals were exposed to welding fumes for several months (50 guinea pigs). Group V–Animals were exposed to welding fumes for 10 months, then infected with TB (10 guinea pigs and 10 controls).

An attenuated strain, the R_1 tubercle bacillus, was used to infect the animals. A water-clear suspension of this strain of tubercle bacilli was prepared from 2-week-old glycerol broth cultures and diluted to give 10 to 15 isolated bacilli per oil immersion field when examined under the microscope. Each animal was infected by force inhalation for 30 seconds of nebulized suspension. All animals reacted strongly to the tuberculin test 6 to 7 weeks after infection. Healing commenced 4 months after infection and was complete in about 12 months.

The results of Group I showed pulmonary alveolar phagocytes filled with brown pigmented particles. The number of phagocytes increased with prolonged exposure, but they did not conglomerate or migrate to lymph nodes (except in rats). Eight guinea pigs died. Chronic bronchitis and tracheitis were noted in all guinea pigs and pneumonitic foci were detected in 39 percent of the animals exposed for 4 to 12 months. Five rats died of chronic pulmonary abscess, a condition that was common in the stock and unrelated to experimental procedures. Four mice died; three of them showed evidence of ordinary chronic murine pneumonia with bronchiectasis.

Animals infected with TB only (Group II) showed discrete subpleural tubercles in the lungs 4 to 5 months after infection. Healing started at about 4 months and tubercles disappeared in about 12 months. Healing usually proceeded by resolution rather than by calcification.

Among the 50 guinea pigs infected with TB after a short exposure to welding fumes (Group III), 9 died within 27 days and 1 died 9-1/2 months later. Examination of the surviving animals at various intervals showed numerous, larger size pulmonary tubercles. Healing proceeded with scar formation and calcification.

Animals exposed to welding fumes during the healing period of TB infection (Group IV) did not exhibit reactivation of the TB. Healing proceeded with calcification rather than by resolution. No guinea pigs died.

The infection of animals exposed to welding fumes for 10 months (Group V) produced a strong positive reaction with the tuberculin test 37 days after infection. The infection was considered light; only a small number of subpleural tubercles were found, and these healed readily. The number, the size, and the healing rate of these tubercles were practically the same in the exposed and the control groups.

The authors concluded that white rats and mice were less susceptible to arc welding gas and fume effects than were guinea pigs. The iron particles inhaled triggered a simple phagocytic response in the lungs but no acute inflammation. Physiological mechanisms, such as interference with phagocytic activity from overloading the lungs with inert mineral particles or, possibly, alterations in the flow of lymph, might have caused the development or larger and more numerous tubercles in animals infected during fume exposure. These lesions healed at the same rate as those of the controls (Group III), never becoming progressive tuberculosis. The welding fumes and gases, furthermore, caused no permanent lung changes in guinea pigs and were found incapable of causing reactivation and spread of tuberculosis from preexisting, partially healed tubercles (Ref. 256).

Exposure to Arc Welding Radiation

Topics including the effects of ultraviolet welding light on vitamin B_1 levels in mice and retinal damage in rabbits and rhesus monkeys from arc welding light are reviewed in this section. The research suggests that only slight degrees of damage to intact eye structures have resulted from exposure to arc welding radiation.

Mice

The effects of UV irradiation on the vitamin B_1 level in mice were studied by Kodama, 1952 (Ref. 161). Fifteen male white mice weighing about 20 g were exposed to UV light (wavelength and intensity not given) for 2 to 6 hours, and the viatmin B_1 content of liver, brain, and eyes was determined. Another group of three mice was irradiated for 2 hours with the animals' eyes closed. Five unexposed mice were used as controls. In addition, 5 mice were placed in a factory where welding was being done for exposure to the UV light generated during welding. Results showed a considerable reduction in vitamin B_1 content of the liver, brain, and eyes of mice exposed to UV light, according to Table D21. The author's conclusion was that large doses of vitamin B_1 should be administered to welders excessively exposed to ultraviolet light, because a vitamin B_1 deficiency can cause reduced powers of accommodation of the eye in humans. A discussion of ultraviolet light effects on the human eye is presented in Chapter 2 of this report.

Rabbits

Clark, 1968 (Ref. 301), noted that the rabbit retina is more susceptible to thermal injury than the human retina, and albino rabbits are ten times less susceptible to thermal retinal damage than ordinary rabbits. Unfortunately, the following study on exposure of rabbits to focused welding arc light failed to specify the strain of rabbit used, although resulting injuries were not severe.

Walther and Szilagy, 1970 (Ref. 301), studied the effects of electric arc welding light on the eyes of rabbits. Nine 4 to 5 kg rabbits (strain not specified) were exposed to arc light focused through a 5 diopter quartz crystal with a focal point of 20 cm. Five electrodes were consumed per animal; each received 12 hours of exposure and was then killed. Eight controls remained unexposed. Histological eye examinations revealed slight corneal opacification, severe keratoconjunctivitis, and a somewhat limited extent of retinal edema. Retinal damage was further ascertained by histochemical enzyme determinations of lactic dehydrogenase, α-glycerophosphate dehydrogenase, β-hydroxybutyrate dehydrogenase, ethanolic dehydrogenase, and nonspecific esterase. Only lactic dehydrogenase (LDH) and nonspecific esterase reacted strongly in controls. In irradiated rabbits, LDH activity was reduced. Thus, the authors suggested that in the absence of firm histological evidence of the early stages of retinal damage, the demonstration of LDH might facilitate detection of this injury.

Table D21
Mean vitamin B_1 levels in mice exposed to ultraviolet lights

Exposure time (hr)	Free vitamin B_1 (μg%)			Total vitamin B_1 (μg%)		
	Liver	Brain	Eyes	Liver	Brain	Eyes
Control	81	66	300	500	355	1,725
2	66	44	123	489	216	1,365
3	40	36	88	450	162	819
4	35	31	46	334	139	600
5	30	12	24	257	133	540
6	23	9	0	195	64	0
2 (eyes closed)	65	63	99	613	169	1,050
2 (factory welding site)	40	2	0	154	57	108

Kodama, 1952 (Ref. 161).

Monkeys

Ham et al., 1976 (Ref. 258), exposed the retinae of 5 rhesus monkeys to eight monochromatic laser lines from 1064 to 441.6 nm. Maximum retinal irradiance and maximum temperature of the retina above ambient were noted for each exposure duration and wavelength. An irradiance of 24 W/cm^2 at 23° C above ambient produced a threshold lesion in 1000 seconds with a 1064 nm laser wavelength. In contrast, the 441.6 nm wavelength required only 30 mW/cm^2 with a negligible temperature rise to produce a threshold lesion in a 1000 second exposure. The increase in retinal sensitivity with a decrease in wavelength was steep, even on a semi-logarithmic plot. The longer wavelength produced a burn, a thermal effect; the shorter wavelength caused a yellowish-white patch with irregular boundaries on the retina, due to photochemical damage. The sensitivity of the retina to blue light (shorter wavelengths) of the welding arc is due to presently unknown photochemical mechanisms. The authors could not distinguish between rod and cone damage or the extent of involvement of retinal photopigments. The lesion seemed to be located in the outer segment of the photoreceptor and possibly in the pigment epithelium. It was noted that although the sensitivity of the retina may continue to increase in the shorter wavelength bands (440 nm), the number of photons reaching the retina is drastically reduced by absorption and scattering in the lens and ocular media.

Glossary of Medical Terms

Most of the definitions contained herein were obtained from *Dorland's Illustrated Medical Dictionary*, 25th Edition, W.B. Saunders Company, Philadelphia, PA, 1974, and are used by permission.

accommodation (of eye). Adjustment for vision at various distances.

actinic (for example, dermatitis). Pertaining to the rays of light beyond the violet end of the spectrum that produce chemical effects.

actinic photophthalmia. Ophthalmia (severe inflammation of the eye or conjunctiva) caused by intense light; for example, rays of a welding arc.

alopecia. Absence of the hair from skin areas where it normally is present.

alveolar. Pertaining to an alveolus, an air cell of the lungs.

andrology. Scientific study of the masculine constitution and of the diseases of the male sex, especially the study of diseases of the male organs of generation.

apical. Pertaining to or located at the apex.

apex (of the lung). The rounded upper extremity of either lung, extending upward as high as the first thoracic vertebra.

arc eye. See **actinic photophthalmia.**

asbestosis. Lung disease (pneumoconiosis) caused by inhaling particles of asbestos.

atherosclerosis. A lesion of large and medium-sized arteries, with deposits in the intima of yellowish plaques containing cholesterol, lipoid material, and lipophages.

atrial fibrillation. A condition characterized by irregular convulsive movements of the atria of the heart, the number of impulses being very great, and individual fibers acting independently.

audiometry. The testing of the sense of hearing.

basophilic stippling. A spotted appearance of red blood corpuscles.

bleb. A bulla or skin vesicle (blister) filled with fluid.

bronchitis. Inflammation of the bronchial tubes.

carboxyhemoglobin. A compound formed from hemoglobin on exposure to carbon monoxide, with formation of a covalent bond with oxygen and without change of the charge of the ferrous state.

carcinoma. A malignant new growth made up of epithelial cells.

catarrh. Inflammation of a mucous membrane with a free discharge.

cerebrovascular. Pertaining to the blood vessels of the cerebrum, or brain.

cholecystitis. Inflammation of the gallbladder.

chorioretinal. Pertaining to the choroid and retina.

choroid or choroidea. The thin, dark brown, vascular coat investing the posterior five-sixths of the eyeball.

ciliary. Pertaining to or resembling the eyelashes or cilia; used particularly in reference to certain structures in the eye, as the ciliary muscle, ciliary process, and ciliary ring.

conjunctiva. The delicate membrane that lines the eyelids and covers the exposed surface of the eyeball.

conjunctivitis. Inflammation of the conjunctiva.

cor pulmonale. Heart disease secondary to disease of the lungs or of their blood vessels.

cornea. The transparent structure forming the anterior part of the fibrous tunic of the eye.

cyanotic. Having a bluish discoloration, especially of the skin and mucous membranes, due to excessive concentrations of reduced hemoglobin of the blood.

cytoplasm. The protoplasm of a cell exclusive of that of the nucleus.

desquamation. The shedding of epithelial elements, chiefly of the skin, in scales or sheets.

diastolic. Of or pertaining to diastole; that is, the period of dilatation of the heart, especially of the ventricles.

dyscrasia (blood). An abnormal or pathologic condition of the blood.

dyspnea. Difficult or labored breathing.

ecchymosis. An extravasation of blood under the skin.

edema. The presence of abnormally large amounts of fluid in the intercellular tissue spaces of the body.

ejaculum. The fluid discharged at ejaculation in the male.

electrocardiogram. A graphic tracing of the electric current produced by the contraction of the heart muscle.

electromyography. The recording of changes in the electrical potential of muscle.

emphysema. A swelling or inflation due to the presence of air, applied especially to a morbid condition of the lungs (pulmonary emphysema).

eosinophilia. The formation and accumulation of an abnormally large number of eosinophils (leukocytes) in the blood.

erythema. A name applied to redness of the skin produced by congestion of the capillaries, which may result from a variety of causes.

expectoration. The act of coughing up and spitting out materials from the lungs, bronchi, and trachea.

fibrotic. Pertaining to or characterized by the formation of fibrous tissue.

fluorosis. A condition resulting from ingestion of excessive amounts of fluorine.

forced vital capacity. The maximal amount of gas that can be expelled from the lung by forceful effort after a maximal inspiration.

fovea (central fovea of the retina). A tiny pit, about one degree wide, in the center of the macula lutea, composed of slim, elongated cones.

fundus oculi. The posterior part, or back of the eye.

galactosuria. Presence of galactose in the urine.

gastritis. Inflammation of the stomach.

gastroduodenitis. Inflammation of the stomach and duodenum.

gastroenteritis. Inflammation of the stomach and intestines.

gingival. Pertaining to the gingivae, or the mucous membrane and supporting tissue that overlies the crowns of unerupted teeth and circles the necks of those that have erupted.

granulocyte. Any cell containing granules, especially a leukocyte containing neutrophil, basophil, or eosinophil granules in its cytoplasm.

hemogram. The blood picture; a written record or a graphic representation of the differential blood count.

histamine. An amine, beta-imidazolylethyl-amine, occurring in all animal and vegetable tissues. It is a powerful dilator of the capillaries and a stimulator of gastric secretion.

hypertension. Abnormally high tension; especially high blood pressure.

intercostal. Situated between the ribs.

intermittent claudication. A complex of symptoms characterized by absence of pain or discomfort in a limb when at rest; the commencement of pain, tension, and weakness after walking is begun; intensification of the condition until walking becomes impossible; and the disappearance of the symptoms after a period of rest.

interstitial. Pertaining to or situated in the interstices of a tissue.

ischemia. Deficiency of blood in a part, due to functional constriction or actual obstruction of a blood vessel.

keratoconjunctivitis. Inflammation of the cornea and conjunctiva.

keratosis. A horny growth or a condition attended by horny growths.

lacrimation. The secretion and discharge of tears.

laryngospasm. Spasmodic closure of the larynx.

lenticular opacities. Opaque areas in the lens of the eye.

leukocytosis. An increase in the number of leukocytes in the blood.

leukopenia. Reduction in the number of leukocytes in the blood, the number being 5,000 or less.

lupus erythematosus, systemic. A morbid condition, ranging from mild to fulminating, associated with visceral lesions and characterized by skin eruptions, prolonged fever, and other constitutional symptoms.

lymphocytosis. Excess of normal lymphocytes in the blood or in any effusion.

macula lutea (macula retinae). An irregular yellowish

depression on the retina, about 3 degrees wide, lateral to and slightly below the optic disk; it is the site of absorption of wavelengths of light.

manganism. A toxic condition occurring in persons who work with manganese.

mesothelioma. A tumor developed from mesothelial tissue.

methemoglobinemia. The presence of methemoglobin in the blood.

methemoglobin. A compound formed from hemoglobin by oxidation of the ferrous to the ferric state.

mucopurulent. Containing both mucus and pus.

myelogram. A roentgenogram of the spinal cord.

nephritis. Inflammation of the kidney; a diffuse progressive degenerative or proliferative lesion affecting in various proportions the renal parenchyma, the interstitial tissue, and the renal vascular system.

normochromic. Having a normal color; having a normal hemoglobin content.

peribronchial. Situated around a bronchus.

peritoneum. The serous membrane lining the abdominopelvic walls and investing the viscera.

petechia. A small, round pinpoint, nonraised, perfectly round, purplish red spot caused by intradermal or submucous hemorrhage, which later turns blue or yellow.

pharyngitis. Inflammation of the pharynx.

photophobia. Abnormal intolerance of light.

plumbism. A chronic form of poisoning produced by the absorption of lead or one of the salts of lead.

pneumoconiosis. Dust in the lungs.

pneumonitis. A condition of localized acute inflammation of the lung without gross toxemia; benign pneumonia.

photodermatitis. An abnormal state of the skin in which light is an important causative factor.

porphyria. A disturbance of porphyrin metabolism, characterized by marked increase in formation and excretion of porphyrins or their precursors.

pruritic. Pertaining to or characterized by pruritus or itching.

pulmonary alveoli. Small outpocketings of the alveolar sacs, through whose walls the gaseous exchange takes place.

purpura. A condition characterized by the presence of confluent petechiae or confluent ecchymoses over any part of the body.

residual volume. The amount of gas remaining in the lung at the end of a maximal expiration.

rhinitis. Inflammation of the mucous membrane of the nose.

scleroderma. A systemic disease which may involve the connective tissues of any part of the body, including the skin, heart, esophagus, kidney, and lung.

scotoma. An area of depressed vision within the visual field, surrounded by an area of less depressed or of normal vision.

siderosis. Pneumoconiosis due to the inhalation of iron or other metallic particles; or the deposit of iron in a tissue.

silicosis. Pneumoconiosis due to the dust of stone, sand, or flint containing silicon dioxide.

spermatogenesis. The process of formation of spermatozoa, including spermatocytogenesis and spermiogenesis.

spirogram. A tracing or graph of respiratory movements.

squamous cell carcinoma. Carcinoma developed from squamous epithelium, and having cuboid cells.

subclinical. Without clinical manifestations; said of the early stages or a slight degree of a disease.

systolic. Pertaining to the period of contraction of the heart, especially of the ventricles.

telangiectasis. The spot formed on the skin by a dilated capillary or terminal artery.

tympanic membrane. The membrane separating the middle ear from the external ear; ear drum.

urticaria. A vascular reaction of the skin marked by the transient appearance of smooth, slightly elevated patches, which are redder or paler than the surrounding skin and often attended by severe itching.

vasculature. The vascular system of the body or any part of it.

ventricular fibrillation. A condition characterized by fibrillary twitching of the ventricular muscle, the impulses traversing the ventricles so rapidly that coordinated contractions cannot occur.

References

1. U.S. Bureau of the Census, "Census of Population, Subject Reports, Occupational Characteristics," U.S. Gov't. Printing Office, Washington, DC, 1970.

2. Sosnin, H. A., Personal Communication, 1978.

3. Dreessen, W. C., Brinton, H. P., Keenan, R. V., Thomas, T. R., Place, E. H., and Fuller, J. E., "Health of Arc Welders in Steel Ship Construction," Pub. Health Bull. No. 298, 1947, 200 p.

4. Garnuszewski, Z., and Dobrzynski, W., "Siderotic Pneumoconiosis Co-Existing with Silicosis in Ship Welders," Gruzlica, 32, 687-694 (1964).

5. Dobrzynski, W., "Frequency of Pulmonary Pneumoconiosis in Electric Hull Welders," Rocz. Pomor. Akad. Med. Szczecinie, 10, 277-294 (1974).

6. Kleinfeld, M. J., "Industrial Pulmonary Disease: Clinical and Experimental Observations," Trans. N.Y. Acad. Sci., 32(1/2), 107-126 (1970).

7. Antti-Poika, M., Hassi, J., and Pyy, L., "Respiratory Diseases in Arc Welders," Int. Arch. Occup. Environ. Health, 40(3), 225-230 (1977).

8. Peters, J. M., Murphy, R. L. H., Ferris, B. G., Burgess, W. A., Ranadive, M. V., and Pendergrass, H. P., "Pulmonary Function in Shipyard Welders," Arch. Environ. Health, 26(1), 28-31 (1973).

9. Hunnicutt, T. N., Cracovaner, D. J., and Myles, J. T., "Spirometric Measurements in Welders," Arch. Environ. Health, 8(5), 661-669 (1964).

10. Kleinfeld, M., Messite, J., Kooyman, O., and Shapiro, J., "Welders' Siderosis. A Clinical Roentgenographic, and Physiological Study," Arch. Environ. Health, 19, 70-73 (1969).

11. Barhad, B., Teculescu, D., and Craciun, O., "Respiratory Symptoms, Chronic Bronchitis, and Ventilatory Function in Shipyard Welders," Int. Arch. Occup. Environ. Health, 36(2), 137-150 (1975).

12. Fogh, A., Frost, J., and Georg, J., "Respiratory Symptoms and Pulmonary Function in Welders," Ann. Occup. Hyg., 12(4), 213-218 (1969).

13. Fogh, A., Frost, J., and Georg, J., "Report on a Danish Investigation into the Health and Working Environment of Arc Welders," International Institute of Welding, Commission VIII Document 440-71, 1971.

14. Fogh, A., Frost, J., and Georg, J., "Chronic Bronchitis and Forced Expiratory Volume in Welders," International Institute of Welding, Commission VIII Document 362-69, 1969.

15. Ross, D. S., "Welders' Health – the Respiratory System, and Welding," Metal Construct., 10(3), 119, 121 (1978).

16. Wilhelmsen, L., Oxhoj, H., Bake, B., Rindby, A., and Selin, E., "Investigations on Effects of Electric Arc Welding in a Swedish Shipyard. A Preliminary Report," Bull. Inst. Marit. Trop. Med. Gdynia, 28(1/2), 45-52 (1977).

17. Ulrich, L., Malik, A. E., and Juck, A., "Injurious Effects of Welding on Health," International Institute of Welding, Commission VIII Document 577-74, 1974.

18. Weiss, W., Private Communication.

19. Erman, M. I., and Rappoport, M. B., "Effect Produced on the Body of the Aerosol Formed during Electric Welding in Carbon Dioxide," Gig. Sanit., 35(7), 35-39 (1970).

20. Guskova, N. I., and Komovnikov, G. S., "Effect of Welding Rod Coating Dust on Alveolar Phagocytosis and Collagen Content in the Lungs," Gig. Tr. Prof. Zabol., 6, 12-15 (1974).

21. Samoilova, L. M., and Kireev, V. I., "Morphological Changes in the Lungs of Animals under the Influence of an Aerosol Formed during Welding and Surfacing of Cast Iron," Gig. Tr. Prof. Zabol., (4), 53-54 (1975).

22. Vlasova-Pryadilova, N. V., "An Occupational Health Study of Arc Welding and Cutting of Copper," Gig. Tr. Prof. Zabol., 15(3), 20-24 (1971).

23. Arutyunov, V. D., Batsura, Y. D., Kireev, V. I., and Likachev, Y. P., "Development of Pulmonary Alveolar Proteinosis Following Inhalation of some Industrial Aerosols," Gig. Tr. Prof. Zabol., (9), 33-37 (1976).

24. Naumenko, I. M., "The Effect of Electric-Welding Aerosol on the Concentration of Collagen and Ascorbic Acid in Lung Tissue," Vrach Delo, 1, 92-94 (1966).

25. Kysela, B., Skoda, V., and Holusa, R., "Studies of the Fibrogenicity of Electric Welding Aerosols in an Animal Experiment," Prac. Lek., 25(10), 417-421 (1973).

26. Senczuk, W., and Nater, B., "Hydroxyproline and Proline in the Lungs of Animals Exposed to Welding Dusts," Ann. Pharm., 8, 103-110 (1970).

27. Garnuszewski, Z., and Dobrzynski, W., "Histological Appearance of the Lungs in Experimental Animals Exposed to Welding Fumes," Gruzlica, 34(1), 4-14 (1966).

28. Lunau, F. W., "Ozone in Arc Welding," Ann. Occup. Hyg., 10, 175-188 (1967).

29. American Society for Testing and Materials, "Standard Practice for Safety and Health Requirements Relating to Occupational Exposure to Ozone," ANSI/ASTM E 591-77.

30. U.S. National Institute of Occupational Safety and Health, "Occupational Mortality in Washington State, 1950-1971," Research Reports Vol. I and III, U.S. DHEW Publication No. (NIOSH) 76-175 A and C, April 1976.

31. Ott, M. G., Holder, B. B., and Langner, R. R., "Determinants of Mortality in an Industrial Population," J. Occup. Med., 18(3), 171-177 (1976).

32. Menck, H. R., and Henderson, B. E., "Occupational Differences in Rates of Lung Cancer," J. Occup. Med., 18(12), 797-801 (1976).

33. Breslow, L., Hoaglin, L., Rasmussen, G., and Abrams, H. K., "Occupations and Cigarette Smoking as Factors in Lung Cancer," Am. J. Public Health, 44, 171-181 (1954).

34. Dunn, J. E., Linden, G., and Breslow, L., "Lung Cancer Mortality Experience of Men in Certain Occupations in California," Am. J. Public Health, 50, 1475-1487 (1960).

35. Dunn, J. E., and Weir, J. M., "A Prospective Study of Mortality of Several Occupational Groups. Special Emphasis on Lung Cancer," Arch. Environ. Health, 17(1), 71-76 (1968).

36. Challen, P. J. R., Hickish, D. E., and Bedford, J., "An Investigation of Some Health Hazards in an Inert-Gas Tungsten-Arc Welding Shop," Br. J. Ind. Med., 15, 276-282 (1958).

37. Anon., "Eye Protection Against Indirect Exposure to Arc Welding," J. Am. Med. Assoc., 221, 1171-1172 (1972).

38. Tengroth, B., "Safety Glasses in Welding," Environ. Res., 11(2), 283-284 (1976).

39. Tengroth, B., and Vulcan, J., "Welding Light," Strahlenther., 153(4), 267-272 (1977).

40. Ross, D. S., and Bamber, L., "Welding Injury Accidents," Metal Constr. Br. Weld. J., 5(7), 261-263 (1973).

41. Key, M., Henschel, A., Butler, J., Ligo, R., and Tabershaw, I., eds., "Occupational Diseases. A Guide to their Recognition," U.S. DHEW Publication (NIOSH) 77-181, 1977.

42. English, W. P., "Eye Protection for Welders," Am. Soc. Safety Eng. J., 18, 39-43 (1973).

43. Rieke, F. E., "Arc Flash Conjunctivitis: Actinic Conjunctivitis from Electric Welding Arc," J. Am. Med. Assoc., 122(11), 734-736 (1943).

44. Cascini, F., Gazzerro, F., and Sica, G., "Radiations from Welding and the Visual Apparatus," Folia Med., 49(11), 834-852 (1966).

45. Benson, C. E., "Treatment of Photophthalmia Following Exposure to Rays of Welding Arc," U.S. Nav. Med. Bull., 41, 737-743 (1943).

46. Edbrooke, C. M., and Edwards, C., "Industrial Radiation Cataract: The Hazards and the Protective Measures," Ann. Occup. Hyg., 10, 293-304 (1967).

47. Naidoff, M. A., and Sliney, D. H., "Retinal Injury from a Welding Arc," Am. J. Ophthalmol., 77(5), 663-668 (1974).

48. Ross, D. S., "Welders' Health – Non-Respiratory Aspects," Metal Constr., 10(5), 204, 206, 208, 209 (1978).

49. Marchand, M., Jacob, M., and Lefebvre, J., "The Pneumopathy of Arc Welders. Investigation in 402 Arc Welders Having More Than 5 Years of Experience," Lille Med., 9(2), 139-145 (1964).

50. Franco, G., Genco, F., and Taccola, A., "On the Value of the P Wave – PQ Segment Ratio or Macruz Index in the Electrocardiograms of a Group of Welders," Med. Lav. 64(11), 447-450 (1973).

51. Taccola, A., and Franco, G., "Macruz Index in the E.C.G. of Workers Exposed to the Hazards of Welding," Med. Lav., 66(6), 663-669 (1975).

52. Ross, D. S., and Hewitt, P. J., "Welding Fume and Welders' Health," Occup. Health, 28(11), 520-527 (1976).

53. Sevcik, M., Chalupa, B., Klhuvkova, K., and Hrazdira, C. L., "The Results of Medical Examinations in Arc Welders," Prac. Lek., 12(5), 229-235 (1960).

54. De Walden-Galuszko, K., and Kruminis-Lozowski, J., "Some Psychical and Vegetative Reactions in Shipyard Workers Exposed to the Action of Noxious Chemical Compounds," Bull. Inst. Marit. Trop. Med. Gdynia, 26(3/4), 337-351 (1975).

55. Anderson, E. W., Andelman, R. J., Strauch, J. M., Fortunin, N. J., and Knelson, J. H., "Effect of Low-level Carbon Monoxide Exposure on Onset and Duration of Angina Pectoris, A Study in Ten Patients with Ischemic Heart Disease," Ann. Intern. Med., 79, 46-50 (1973).

56. Wald, N., and Howard, S., "Smoking, Carbon Monoxide and Arterial Disease," Ann. Occup. Hyg., 18, 1-14 (1975).

57. Aronow, W. S., Stemmer, E. R., Wood, B., Zweig, S., Tsao, K., and Raggio, L., "Carbon Monoxide and Ventricular Fibrillation. Threshold in Dogs with Acute Myocardial Injury," Am. Heart J., 95(6), 754-756 (1978).

58. Schuler, P., Maturama, V., Cruz, E., Guijon, C., Vasquez, A., Valenzuela, A., and Silva, R., "Arc Welder's Pulmonary Siderosis," J. Occup. Med., 4(7), 353-358 (1962).

59. Hamilton, A., and Hardy, H. L., "Industrial Toxicology," 3rd ed., Publishing Sciences Group, Inc., Acton, MA, 1974.

60. Chmielewski, J., Jaremin, B., Bartnicki, C., and Konieczka, R., "Evaluation of Occupational Exposure to Zinc Oxide in the Marine Production Shipyard. II. Examination of the State of Health of the Workers Exposed to Zinc Oxide," Bull. Inst. Med. Gdansk, 25(1), 53-65 (1974).

61. Tola, S., Kilpio, J., Virtamo, M., and Haapa, K., "Urinary Chromium as an Indicator of the Exposure of Welders to Chromium," Scand. J. Work Environ. Health, 3, 192-202 (1977).

62. Rozera, G., Serra, P., and Vitacca, L., "Clinical Statistical Contribution to the Pathology of Welders," Folia Med., 49(6), 447-454 (1966).

63. Mazza, V., and Brancaccio, A., "Adrenal Function in Welders," Folia Med., 51(9), 715-719 (1968).

64. Challen, P. J. R., "Basic Electrodes," International Institute of Welding, Commission VIII Document 332-68, 1968.

65. Bahn, A., Private Communication.

66. U.S. National Institute of Occupational Safety and Health, "Criteria for a Recommended Standard. Occupational Exposure to Chromium (VI)," HEW Publication No. (NIOSH) 76-129, 1975.

67. U.S. National Institute of Occupational Safety and Health, "Criteria for a Recommended Standard. Occupational Exposure to Inorganic Nickel," DHEW (NIOSH) Publication No. 77-164, 1977.

68. DeCoufle, P., Private Communication.

69. U.S. National Institute of Occupational Safety and Health, "Criteria for a Recommended Standard, Occupational Exposure to Crystalline Silica." HEW Publication No. (NIOSH) 75-120.

70. American Welding Society, "Fumes and Gases in the Welding Environment," American Welding Society, Miami, FL (1979), 232 p.

71. Buckup, H., "The Welder's Lung: Harmful or Indifferent Pulmonary Changes," Z. Arbeitsmed. Arbeitsschutz, 23(9), 286-293 (1973).

72. Vitums, V. C., Niles, N. R., Borman, J. O., and Lowry, R. D., "Pulmonary Fibrosis from Amorphous Silica Dust, a Product of Silica Vapor," Arch. Environ. Health, 32(2), 62-68 (1977).

73. Styles, J. A., and Wilson, J., "Comparison between *In Vitro* Toxicity of Polymer and Mineral Dusts and their Fibrogenicity," Ann. Occup. Hyg., 16, 241-250 (1973).

74. Klosterkotter, W., and Robock, K., "New Aspects on Dust and Pneumoconiosis Research," Am. Ind. Hyg. Assoc. J., 36(9), 659-668 (1975).

75. Wintrobe, M. W., Thorn, G. W., Adams, R. D., Braunwald, E., Isselbacher, K. J., and Petersdorf, R. G., eds, "Harrison's Principles of Internal Medicine," 7th ed., McGraw-Hill Book Co., New York (1974).

76. Berky, J., and Sherrod, C., eds, "*In Vitro* Toxicity Testing, 1975-1976," Franklin Institute Press, Philadelphia, PA (1977).

77. McCann, J., Choi, E., Yamasaki, E., and Ames, B. N., "Detection of Carcinogens as Mutagens in the Salmonella/Microsome Test: Assay of 300 Chemicals," Proc. Nat. Acad. Sci. USA. 72(12), 5135-5139 (1975).

78. Stern, R. M., "A Chemical, Physical and Biological Assay of Welding Fumes," The Danish Welding Institute Publication 77.05, 1977.

79. American Welding Society Subcommittee on Definitions, "Classification of the Welding Process," Weld, J., 55(4), 270-272 (1976).

80. Moreton, J., "The Generation and Control of Welding Fume. Bibliography: Particulate Fume and Pollutant Gases in the Welding Environment," The Welding Institute Publication 5510/19/77, Cambridge, Eng. (1977), 74 p.

81. Heile, R. F., and Hill, D. C., "Particulate Fume Generation in Arc Welding Processes," Weld. J., 54(7), 201s-210s (1975).

82. American Welding Society, "The Welding Environment," American Welding Society, Miami, FL (1973), 169 p.

83. Maxild, J., Andersen, M., Kiel, P., and Stern, R. M., "Mutagenicity of Fume Particles from Metal Arc Welding on Stainless Steel in the Salmonella/Microsome Test," Mutat. Res., 56(3), 235-243 (1978).

84. Virtamo, M., and Tuomola, S., "Hexavalent Chromium Compounds in Welding Fumes," International Institute of Welding, Commission VIII Document 584-74, 1974.

85. Clapp, D. E., and Owen, R. J., "An Investigation of Potential Health Hazards of Arc Welding Fume Growth with Time," Weld. J., 56(12), 380s-385s (1977)

86. Akselsson, K. R., Desaedeleer, G. G., Johansson, T. B., and Winchester, J. W., "Particle Size Distribution and Human Respiratory Deposition of Trace Metals in Indoor Work Environments," Ann. Occup. Hyg., 19(3/4), 225-238 (1976).

87. Jarnuszkiewicz, I., Knapik, A., and Kwiatkowski, B., "Particle Size Distribution of Welding Fume Depending on the Microclimatic Conditions of the Welders Working Places," Bull. Inst. Mar. Med. Gdansk., 17(1/2), 73-76 (1966).

88. Hedenstedt, A., Jenssen, D., Lidesten, B., Ramel, C., Rannug, U., and Stern, R. M., "Mutagenicity of Fume Particles from Stainless Steel. Welding," Scand. J. Work Environ. Health, 3, 203-211 (1977).

89. Frant, R., "Formation of Ozone in Gas-Shielded Welding," Ann. Occup. Hyg., 6, 113-125 (1963).

90. Pattee, H. E., Myers, L. B., Evans, R. M., and Monroe, R. E., "Effects of Arc Radiation and Heat on Welders," Weld. J., 52(5), 297-308 (1973).

91. Vorontsova, E. I., Karacharov, T. S., and Voschanov, K., "Working Conditions When Arc Welding Aluminum and Aluminum Alloys," Weld. Prod., 7(9), 59-65 (1961).

92. Alpaugh, E. L., and Phillippo, K. A., and Pulsifer, H. C., "Clearing the Fog about Mig Fumes," Weld. Eng., 53(10), 41-45 (1968).

93. Steel, J., "Respiratory Hazards in Shipbuilding and Shiprepairing," Ann. Occup. Hyg., 11, 115-121 (1968).

94. Doig, A. T., and Challen, P. J. R., "Respiratory Hazards in Welding," Ann. Occup. Hyg., 7, 223-231 (1964).

95. Mourik, J. van, "The Formation of Nitrous Fumes in Gas Flames," Ann. Occup. Hyg., 10, 305-315 (1967).

96. Emmett, E. A., and Horstman, S. W., "Factors Influencing the Output of Ultraviolet Radiation During Welding," J. Occup. Med., 18(1), 41-44 (1976).

97. Lyon, T. L., Marshall, W. J., Sliney, D. H., Krial, N. P., and Del Valle, P. F., "Evaluation of the Potential Hazards from Actinic Ultraviolet Radiation Generated by Electric Welding and Cutting Arcs," National Technical Information Service Report AD/A-033 768, 1976.

98. American Welding Society, "Arc Welding and Cutting Noise," American Welding Society, Miami, FL (1979), 39 p.

99. Damborenea, L. L., "Incomplete History of a Clinical Case (Pathology of the Voltaic Arc Syndrome)," Rev. Espan. Anest. Rean., 23, 502-505 (1976).

100. Glass, W. I., Harris, E. A., and Whitlock, R. M., "Phosgene Poisoning: Case Report," New Zealand Med. J., 74(475), 386-389 (1971).

101. Montesano, G., "Acute Pulmonary Syndrome from Arc Welding," Med. Lav., 41(8/9), 225-229 (1950).

102. U.S. National Institute of Occupational Safety and Health, "Criteria for a Recommended Standard. Occupational Exposure to Phosgene," HEW Publication No. (NIOSH) 76-137.

103. U.S. National Institute of Occupational Safety and Health, "Criteria for a Recommended Standard. Occupational Exposure to Cadmium," HEW Publication No. (NIOSH) 76-192.

104. Wheater, R. H., "Effects of Ozone Produced by Argon-CO_2-Shielded Arc Welding," J. Am. Med. Assoc., 236, 605-607 (1976).

105. Kleinfeld, M., and Giel, C. P., "Clinical Manifestations of Ozone Poisoning: Report of a New Source of Exposure," Am. J. Med. Sci., 231, 638-643 (1956).

106. Moles, J. E., and Collins, R. J., "Ozone Exposure from Arc Welding," Trans. 19th Ann. Mtg. ACGIH,

April 20-23, 1957, pp. 33-36, NIOSH, April 23, 1957.

107. Kurta, A. V., "Acute Pulmonary Edema Due to Welding in a Confined Space," Austral. New Zealand J. Med., 6(4), 369 (1976).

108. Kleinfeld, M., Giel, G., and Tabershaw, I. R., "Health Hazards Associated with Inert-Gas-Shielded Metal Arc Welding," A.M.A. Arch. Ind. Health, 15, 27-31 (1957).

109. U.S. National Institute of Occupational Safety and Health, "Recommended Standard for Occupational Exposure to Oxides of Nitrogen," U.S. Government Printing Office: 1976-657-645/247, (1976).

110. Morley, R., and Silk, S. J., "The Industrial Hazard from Nitrous Fumes," Ann. Occup. Hyg., 13, 101-107 (1970).

111. Blejer, H. P., "Death Due to Cadmium Oxide Fumes," Ind. Med. Surg., 35(5), 363-364 (1966).

112. Axford, A. T., "Industrial Diseases of the Lung," International Institute of Welding, Commission VIII Document 612-75, 1975.

113. Blejer, H. P., Caplan, P. E., and Alcocer, A. E., "Acute Cadmium Fume Poisoning in Welders – A Fatal and a Nonfatal Case in California," Calif. Med., 105(4), 290-296 (1966).

114. Beton, D. C., Andrews, G. S., Davies, H. J., Howells, L., and Smith, G. F., "Acute Cadmium Fume Poisoning," Br. J. Ind. Med., 23(4), 292-301 (1966).

115. Christensen, F. C., and Olson, E. C., "Cadmium Poisoning," A.M.A. Arch. Ind. Health, 16, 8-13 (1957).

116. Patwardhan, J. R., and Finckh, E. S., "Fatal Cadmium-Fume Pneumonitis," Med. J. Austral., 1(25), 962-964 (1976).

117. Townshend, R. H., "A Case of Acute Cadmium Pneumonitis: Lung Function Tests During a Four-Year Follow-Up," Br. J. Ind. Med., 25(1), 68-71 (1968).

118. Clack, G., "Shipyard Health Hazards," Job Safety *Health, 3(3), 5-10 (1975).*

119. Collen, M. F., "A Study of Pneumonia in Shipyard Workers with Special Reference to Welders," J. Ind. Hyg. Toxicol., 29(2), 113-121 (1947).

120. Poinso, R., Rouyer, P., and Jouve, G., "Pulmonary Changes in Electric Arc Welders," Arch. Mal. Prof., 14, 479-487 (1953).

121. Charr, R., "Respiratory Disorders Among Welders," Am. Rev. Tubercul. Pulmon. Dis., 71, 877-884 (1955).

122. Garnuszewski, Z., and Dobrzynski, W., "Histological Picture of the Bronchial Mucosa in Welders Employed at the Warski Shipyard in Szczecin," Gruzlica, 34(1), 15-17 (1966).

123. Stettler, L. E., Groth, D. H., and Mackay, G. R., "Identification of Stainless Steel Welding Fume Particles in Human Lung and Environmental Samples Using Electron Probe Microanalysis," Am. Ind. Hyg. Assoc. J., 38(2), 76-82 (1977).

124. Angervall, L., Hansson, G., and Rockert, H., "Pulmonary Siderosis in an Electric Welder," Acta. Pathol. Microbiol. Scand., 49(3), 373-380 (1960).

125. Morgan, W., and Kerr, H. D., "Pathologic and Physiologic Studies of Welders' Siderosis," Ann. Intern. Med., 58(2), 293-304 (1963).

126. Brun, J., Cassan, G., Kofman, J., Gilly, J., and Magnin, F., "Conglomerate Siderosis of Arc Welders," J. Fr. Med. Chir. Thorac., 26(2), 133-142 (1972).

127. Brun, J., Cassan, G., Kofman, J., and Gilly, J., "Siderosclerosis of Arc Welders with Diffuse Interstitial Fibrosis and a Pseudo-Tumoral Conglomerative Picture," Poumon Coeur, 28(1), 3-10 (1972).

128. Cassan, G., Kofman, J., Gilly, J., and Brun, J., "Pulmonary Siderosis in Electric Welders: Diffuse Interstitial Fibrosis with Conglomerative Form," Arch. Mal. Prof. Med. Trav. Secur. Soc., 33, 193-195 (1972).

129. Spacilova, M., and Koval, Z., "Pulmonary X-ray and Functional Findings in Electric-Arc Welders," Int. Arch. Arbeitsmed., 34(3), 231-236 (1975).

130. Groh, J. A., "Benign Pulmonary Changes in Arc-Welders – Arc Welder's Siderosis," Ind. Med., 13(8), 598-601 (1944).

131. Britton, J. A., and Walsh, E. L., "Health Hazards of Electric and Gas Welding," J. Ind. Hyg. Toxicol., 22, 125-151 (1940).

132. Kierst, W., Uselis, J., Graczyk, M., and Krynicki, A., "Pulmonary Changes in Shipyard Arc-Welders," Bull. Inst. Mar. Med. Gdansk, 15(3), 149-156 (1964).

133. Gillon, J. J., and Marchand, M., "A Follow-up Study on the Health of 500 Arc Welders after more than 5 Years at this Work," in: XIVth International Congress of Occupational Health, Vol. III, Free Communications, p. 1125-1126, 1963, Madrid, Spain, 16-21 Sept., (1963).

134. Harding, H. E., McLaughlin, A. I. G., and Doig, A. T., "Clinical, Radiographic, and Pathological Studies of the Lungs of Electric-Arc and Oxy-acetylene Welders," Lancet, 3, 394-400 (1958).

135. Mann, B. T., and Lecutier, E. R., "Arc Welders' Lung," Br. Med. J., 2, 921-922 (1957).

136. Friede, E., and Rachow, D. O., "Symptomatic Pulmonary Disease in Arc Welders," Ann. Internal Med., 54, 121-127 (1961).

137. Meyer, E. C., Kratzinger, S. F., and Miller, W. H., "Pulmonary Fibrosis in an Arc Welder," Arch. Environ. Health, 15(4), 462-469 (1967).

138. Morgan, W. K., "Arc Welders' Lung Complicated by Conglomeration," Am. Rev. Resp. Dis., 85, 570-575 (1962).

139. Benavides, R., "A Clinical Radiological and Functional Study in Arc Welders and Oxyacetylene Cutters," Rev. Med. Chil., 104(9), 610-614 (1976).

140. Stanescu, D. C., Pilat, L., Gavrilescu, N., Teculescu, D. B., and Cristescu, I., "Aspects of Pulmonary Mechanics in Arc Welders' Siderosis," Br. J. Ind. Med., 24(2), 143-147 (1967).

141. Ross, D. S., "Welders' Metal Fume Fever," J. Soc. Occup. Med., 24(4), 125-129 (1974).

142. Lange, A., Smolik, R., Zatonski, W., Zwolinski, J., Garncarek, D., Okrojek, M., and Gielgier, Z., "Serum Immunoglobulin and Other Protein Levels in Arc Welders," Int. Arch. Occup. Environ. Health, 38(3), 189-196 (1977).

143. Lotti, M., and Mazzotta, M., "Alpha-1-Antitrypsin, STIC (Serum Trypsin Inhibitory Capacity) and TPE (Trypsin-Protein Esterase) Activity in the Serum of a Group of Welders," Med. Lav., 64(11/12), 437-443 (1973).

144. Stumphius, J., and Meyer, P. B., "Asbestos Bodies and Mesothelioma," Ann. Occup. Hyg., 11(4), 283-293 (1968).

145. Houten, L., and Bross, I. D. J., "Occupational Cancer in Men Exposed to Metals," Adv. Exp. Med. Biol., 91, 93-102 (1977).

146. Powell, C. H., Goldman, L., and Key, M. M., "Investigative Studies of Plasma Torch Hazards," Am. Ind. Hyg. Assoc. J., 29(4), 381-385 (1968).

147. Hinrichs, J. F., "Project Committee on Radiation – Summary Report," Weld. J., 57(1), 62, 65 (1978).

148. Haneke, E., and Gutschmidt, E., "Squamous Cell Carcinoma in an Electric Arc Welder," Berufsdermat., 24(5), 119-122 (1976).

149. Roquet-Doffiny, Y., Roquet, P., and La Chapelle, J. M., "Burns, Actinic Dermatitis and Multiple Epitheliomas in Welders," Dermatologica, 155(3), 176-179 (1977).

150. Bates, C. C., "The Effects on the Human Eye of the Radiant Energy Given Off by Various Welding Processes," Sheet Metal Ind., 29(300), 349-352, 357; 29(301), 435-440, 447 (1962).

151. Entwistle, H., "A Case-Book of Welding Accidents," Ann. Occup. Hyg., 7, 207-221 (1964).

152. Shiras, F., "Radiant Energy Eye Protection Devices, Part I: Welding Operations, Part II: Molten Metal Glare," Nat. Safety News, 93(3), 63-70 (1966).

153. Sliney, D. H., and Freasier, B. C., "Evaluation of Optical Radiation Hazards," Appl. Optics, 12(1), 1-24 (1973).

154. Pitts, D. G., and Tredici, T. J., "The Effects of Ultraviolet on the Eye," Am. Ind. Hyg. Assoc. J., 32, 235-246 (1971).

155. Ross, D. S., "Welders' Health – A Review of Currently Available Information," Metal. Constr., 9(10), 475-479 (1977).

156. El Gammal, M. Y., Soliman, A. M., and Mostafa, M. S., "Actinic Conjunctivitis. The Effect of Ultraviolet and Infrared Radiation on the Eyes of Welders and Glass Blowers," Bull. Ophth. Soc. Egypt, 66(70), 41-52 (1973).

157. Verin, P., and Peyrat, M., "Protection of Eyes of Welders," Bull. Soc. Ophthalmol. Fr., 72(9/10), 893-895 (1972).

158. Abrahamson, I. A., "Arc Welders Burn," Am. Family Physic., 13(2), 145 (1976).

159. Sykowski, P., "Dendritic Type of Keratitis Caused by a Welding Arc," Am. J. Ophthalmol., 34, 99-100 (1951).

160. Szafran, L., "Lens Opacity of the Smelting Cataract Type in Welders," Med. Pr., 16(3), 246-249 (1965).

161. Kodama, M., "Vision and Vitamin B_1 Metabolism in Arc Welders," Acta Soc. Ophthalmol. Japan, 56, 453-462 (1952).

162. Wheater, R. H., "Eye Damage from Repeated Exposure to Welding Arcs," J. Am. Med. Assoc., 236, 2224 (1976).

163. Filipiakowa, Z., "The Influence of Welding Arc Radiation on the Picture of the Eye Fundus," Klin. Oczna, 40, 529-533 (1970).

164. Ruprecht, K. W., "Foveal Maculopathy due to Electric Arc Welding," Zentralbl. Arbeitsmed., 26(10), 220-223 (1976).

165. Guyton, A. C.,"Textbook of Medical Physiology," 4th ed., W. B. Saunders Co., Philadelphia, PA, 1971.

166. Filipiakowa, Z., "Color Vision in Electric Welders," Klin. Oczna, 45(8), 933-938 (1975).

167. Filipiakowa, Z., "The Influence of the Radiation Energy of the Welding Arc on Dark Adaptation," Klin. Oczna, 41(1), 49-55 (1971).

168. Petersen, I., Ekberg, K., Kadefors, R., and Persson, J., "Experiences in Clinical Neurophysiology," Bull. Inst. Marit. Trop. Med. Gdynia, 28(1/2), 73-82 (1977).

169. Ross, D. S., and Parkes, L. R., "Arc Welding – The Health Aspect," Metallurg. Materials Technol., 7(4), 191-194 (1975).

170. Ross, D. S., "Loss of Consciousness in a Burner Using an Oxyacetylene Flame in a Confined Space," Ann. Occup. Hyg., 13, 159-160 (1970).

171. Rieke, F. E., "Lead Intoxication in Shipbuilding and Shipscrapping: 1941 to 1968," Arch. Environ. Health, 19(4), 521-539 (1969).

172. National Academy of Sciences, "Manganese," National Academy of Sciences, Washington, DC, 1973, 191 p.

173. Vorontsova, E. I., "Action on the Organism of the Aerosol Obtained in Arc Welding with Manganese Electrodes," Gig. Sanit., (4), 24-30 (1949).

174. Hine, C. H., and Pasi, A., "Manganese Intoxication," West, J. Med., 123(2), 101-107 (1975).

175. Browning, E., "Toxicity of Industrial Metals," 2nd ed., Butterworths, London, 1969, p. 213-225.

176. Beintker, E., "Manganese Effects in Electric Arc Welding," Zentralbl. Gewerbehyg., 9, 207-211 (1932).

177. Platanov, A. A., "Investigation of Some Features of the Personality of Patients with Manganese Poisoning," Gig. Tr. Prof. Zabol., (10), 20-24 (1976).

178. Barborik, M., "Health Hazards in Arc Welding," Acta Univ. Palack. Olumuc. Fac. Med., 46, 69-89 (1967).

179. Pesaresi, C., Cascini, F., Lamanna, P., and Brancaccio, A., "Manganese Risk in Workers Employed in Electric Arc Welding with Covered Electrode," Folia Med., 51(6), 466-477 (1968).

180. Ponomareva, R. I., "Excretion of Manganese in the Urine," Gig. Sanit., 22, 69-70 (1957).

181. Barborik, M., "The Effect of Manganese Compounds on the Health of Welders using Acid Electrodes," Prac. Lek., 25(3), 82-86 (1973).

182. Jindrichova, J., "Manganese in Feces of Welders Practicing Submerged Arc Welding," Prac. Lek., 25(3), 86-90 (1973).

183. Oltramare, M., Tchicaloff, M., Desbaumes, P., and Hermann, G., "Chronic Manganese Poisoning in Two Arc Welders," Int. Archiv. Gewerbepathol. Gewerbehyg., 21, 124-140 (1965).

184. Meneghini, P., and D'Onofrio, V., "Remarks on the Behavior of Hemopoietic Marrow in Arc Welders," Med. Lav., 40(2), 43-48 (1949).

185. Piscator, M., "Health Hazards from Inhalation of Metal Fumes," Environ. Res., 11(2), 268-270 (1976).

186. Bataille, M., "Hemorrhagic Hemopathy in a Welder," Rev. Stomatol., 47(1/2), 15-18 (1946).

187. Haneke, E., "Findings in the Ejaculate of Electro Welders," Dermatol. Monatsschr., 159(11), 1036-1040 (1973).

188. Aronow, W. S., and Cassidy, J., "Effect of Carbon Monoxide on Maximal Treadmill Exercise, A Study in Normal Persons," Ann. Intern. Med., 83, 496-499 (1975).

189. Cohen, S. I., Deane, M., and Goldsmith, J. R., "Carbon Monoxide and Survival from Myocardial Infarction," Arch. Environ. Health, 19, 510-517 (1969).

190. Luhr, J., "Permanent Auricular Fibrillation Due to Poisoning from Nitrous Gases Formed During Welding," Nord. Med., 53, 577-579 (1955).

191. Krivoglaz, B. A., Navakatikian, A. O., Boiko, V. G., Zaritskaia, L. A., and Krasniuk, E. P., "State of Health of Workers Welding in a Carbon Dioxide Environment," Vrach Delo, 2, 140-144 (1973).

192. Harrold, G. C., Meek, S. F., and McCord, C. P., "A Chemical and Physiological Investigation of Electric Arc Welding," J. Ind. Hyg. Toxicol., 22(8), 347-348 (1940).

193. Stancari, V., and Amorati, A., "Gastroduodenal Diseases in Arc Welders," Arch. Pat. Clin. Med., 40, 72-80 (1963).

194. Sklensky, B., "Subacute Lead Poisoning," Vnitrni Lekar., 15(1), 89-92 (1969).

195. Salamone, L., Rizzo, A., and La Paglia, G., "Clinico-Statistical Findings on Chronic Pathology of Welders," Folia Med., 52(8), 511-525 (1969).

196. Baranova, V. M., Keizer, S. A., Smirnov, V. F., Zertsalov, U. V., Klocheva, N. M., Tatarskaya, A. A., Rezanov, P. S., and Sebeleva, V. I., "Pathological Characteristics of Shipbuilding Electrical Welders," Gig. Tr. Prof. Zabol., (6), 45-47 (1976).

197. Schuckmann, F., "General and Specific Health Hazards of Welding," Zentralbl. Arbeitsmed., 25(4), 97-101 (1975).

198. Kny, W., "Welders' Disease," Arch. Gewerbepath.

Gewerbehyg., 11, 369-380 (1942).

199. U.S. National Institute of Occupational Safety and Health, "Criteria for a Recommended Standard. Occupational Exposure to Inorganic Fluorides," U.S. DHEW Publication No. (NIOSH) 76-103, 1975.

200. Pantucek, M. B., "Hygiene Evaluation of Exposure to Fluoride Fume from Basic Arc-Welding Electrodes," Ann. Occup. Hyg., 18(3), 207-212 (1975).

201. Achapovskaya, A. M., "Working Posture of the Operators of Welding Rigs," Gig. Sanit. (9), 102-103 (1976).

202. Kadefors, R., Petersen, I., and Herberts, P., "Muscular Reaction to Welding Work: An Electromyographic Investigation," Ergonomics, 19(5), 543-558 (1976).

203. Petersen, I., Kadefors, R., and Persson, J., "Neurophysiologic Studies of Welders in Shipbuilding Work," Environ. Res., 11(2), 226-236 (1976).

204. Ilner-Paine, O., "Swedish Shipyard Workers Give Health Hazards the Cold Shoulder," New Scientist, 73, 403 (1977).

205. Volfovskaya, R. N., and Makulova, I. D., "Characteristics of Toxic Pneumonia in Electro- and Gas-Welders," Nauchn. Tr. Leningr. Gos. Inst. Usoversh. Vrachei, 98, 145-150 (1971).

206. Fishburn, C. W., and Zenz, C., "Metal Fume Fever. A Report of a Case," J. Occup. Med., 11(3), 142-144 (1969).

207. Paichl, P., Skrlant, L., and Sykora, J., "Fever due to Inhalation of Metal Fumes from Brass Welding," Plzensky Lek. Sbor., 23, 115-118 (1964).

208. Ilic, V., and Popovic, K., "Metal Fume Fever," Arh. Hig. Rada. Toksikol., 22, 359-364 (1971).

209. Molfino, F., "Zinc Poisoning in Electrical and Autogenic Welding Employed in Naval Repair and Construction," Rass. Med. Applicata. Lavoro. Ind., 8, 341-343 (1937).

210. Glass, W. I., "Mercury Fume Fever," New Zealand Med. J., 71(456), 297-298 (1970).

211. Petersen, D. J. van, "Sudden Deafness Caused by a Drop of Welding Metal," Ned. Tijdschr. Geneeskunde, 115(39), 1600-1601 (1971).

212. Mobius, M., "Welding Injury of the Tympanic Membrane," Hals, Nasen, Ohrensaerzte, 12, 18-22 (1964).

213. Moller, P., "Glowing Particles in the Ear," T. Norske Laegeforen, 95(7), 445-446 (1975).

214. Andreasson, L., and Elner, A., "Welding and Ear Injuries," Lakartidningen, 71(25), 2553-2554 (1974).

215. Rudenko, V. F., and Kudrya, V. F., "Hygiene Assessment of Noise during Gas Welding and Cutting Operations and Preventive measures," Weld. Prod., 22(3), 72-73 (1975).

216. Brusin, A., Petrovic, D., Kostic, M., and Dimovic, A., "Acoustic Trauma in Welders Exposed to Overall Noise in Mechanical-Technological Workshops," Med. Pregl., 22(11), 545-549 (1969).

217. Hickish, D. E., and Challen, P. J. R., "A Preliminary Study of Some Health Hazards in the Plasma Jet Process," Br. J. Ind. Med., 20, 95-99 (1963).

218. Balabanow, K., Andreev, V. C., and Botev, N., "A Case of Photoelectric Dermatitis with a Clinical and Histological Course of Lupus Erythematosus," Dermatol. Wochenschr., 153(21), 609-616 (1967).

219. Schmitt, C. L., and Silverman, A., "Discoid Lupus Erythematosus in an Arc Welder," Cutis, 8(5), 476-477 (1971).

220. Gunche, F. F., and Feoli, R. L. S., "Occupational Lupus Erythematosus," Sem. Med., 116, 3-5 (1960).

221. Wozniak, K. D., "Erythematodes Chronicus Discoides as an Occupational Disease in an Electric Arc Welder," Berufsdermat., 19(4), 187-196 (1971).

222. Szarmach, H., and Synoradzka-Nakonieczna, H., "Evaluation of the Effect of Welding on the Occurrence of Dermatoses in Ship Industry," Przegl. Dermatol., 57(2), 195-202 (1970).

223. Hjorth, N., and Wilkinson, D. S., "Contact Dermatitis 8. Ecology of Contact Dermatitis," Br. J. Dermatol., 81(10), 796-798 (1969).

224. Gylseth, B., Gundersen, N., and Langard, S., "Evaluation of Chromium Exposure Based on a Simplified Method for Urinary Chromium Determination," Scand. J. Work Environ. Health, 3(1), 28-31 (1977).

225. Duengemann, H., Borelli, S., and Wittmann, J., "Contact Eczema from Copper and Cadmium in Welders, Grinders, Galvanizers and Similar Occupational Groups," Arbeitsmed. Socialmed. Arbeitshyg., 7(4), 85-93 (1972).

226. Fregert, S., and Ovrum, P., "Chromate in Welding Fumes with Special Reference to Contact Dermatitis," Acta Dermato-Venereol., 43(2), 119-124 (1963).

227. Kaplan, I., and Zeligman, I., "Urticaria and Asthma from Acetylene Welding," Arch. Dermatol., 88, 188-189 (1963).

228. Fessel, W. J., "Scleroderma and Welding," N. Engl. J. Med., 296(26), 1537 (1977).

229. Ulfvarson, U., and Wold, S., "Trace-Element Con-

centrations in Blood Samples from Welders of Stainless Steel or Aluminum and a Reference Group," Scand. J. Work Environ. Health, 3, 183-191 (1977).

230. De Kretser, A. J., Evans, W. D., and Waldron, H. A., "Carbon Monoxide Hazard in the CO_2 Arc – Welding Process," J. Occup. Hyg., 7, 253-259 (1964).

231. American Conference of Governmental Industrial Hygienists, "TLVs. Threshold Limit Values for Chemical Substances and Physical Agents in the Workroom Environment with Intended Changes for 1977," ACGIH, Cincinnati, OH (1977).

232. Krechniak, J., "Fluoride Hazards Among Welders," Fluoride Quart. Rep., 2(1), 13-24 (1969).

233. Jahr, J., and Johnsen, B. K., "Nickel Exposure and Excretion During Welding of Stainless Steel," International Institute of Welding, Commission VIII Document 587-74, 1974.

234. Migai, K. V., and Norkin, Y. N., "The Biological Action of Electric Welding Dusts and Gases formed in the Welding of Special Steels Using Chrome-Nickel Electrodes," Nauch Raboty Inst. Okhrany Tr. Vses. Tsentr Sovet Prof. Soyuzov, 2(34), 45-53 (1965).

235. Von Haam, E., and Groom, J. J., "The Pathology of Shielded Arc Welding," J. Ind. Hygiene Toxicol., 23(2), 55-77 (1941).

236. McCord, C. P., Harrold, G. C., and Meek, S. F., "A Chemical and Physiological Investigation of Electric Arc Welding. III. Coated Welding Rods," J. Ind. Hyg. Toxicol., 23(5), 200-215 (1941).

237. Hewitt, P. J., and Hicks, R., "An Investigation of the Effects of Inhaled Welding Fume in the Rat," Ann. Occup. Hyg., 16(3), 213-221 (1973).

238. Hewitt, P. J., and Hicks, R., "The Retention of Some Metallic Elements in Tissues of Rats Exposed to Welding Fumes," Br. J. Pharmacol., 46(3), 571P-572P (1972).

239. Vorontsova, E. I., Kochetkova, T. A., and Priadilova, N. V., "Changes in Lungs of Animals in Copper Welding," International Institute of Welding, Commission VIII Document 376-69, 1969.

240. Leonicheva, V. D., "Experimental Pneumoconiosis Produced by Dust Forming when Aluminum and Aluminum-Magnesium Alloys are Welded," Gig. Tr. Prof. Zabol., 9(10), 11-16 (1965).

241. Likhachev, I. P. Batsura, I. D., and Direev, V. I., "The Role of Several Occupational Factors in the Development of Pulmonary Alveolar Proteinosis," Arkh. Patol., 37(2), 63-69 (1975).

242. Kellermann, K., and Lehmann, G., "Fumes and Gas in Welding with Basic (Lime-Bearing) Electrodes," Br. Weld. J., 3, 196-200 (1956).

243. Dabrowski, Z., Meyer, J., and Senczuk, W., "Fertility and the Morphologic Changes in the Genital Organs of Rat Females Exposed to the Action of Welding Fumes and Gases," Ginekol. Pol., 37(8), 845-852 (1966).

244. Dabrowski, Z., Meyer, J., and Senczuk, W., "Fecunolity and Morphologic Changes in the Testes of Rats Exposed to Welding Fumes and Gases," Ginekol. Pol., 37(10), 1037-1044 (1966).

245. Byczkowski, S., Cempel, M., and Zakrzewska, T., "Tissue Oxidation in Livers of Rats Exposed to Welding Fumes," Farm. Polska, 21(19-20), 734-738 (1965).

246. Titus, A. C., Warren, H., and Drinker, P., "Electric Welding I. The Respiratory Hazard," J. Ind. Hygiene, 17(4), 121-128 (1935).

247. Kawada, K., Iwano, K., Nagai, N., and Kimura, S., "Experimental Studies on the Harmfulness of Basic Type Welding Electrodes and on their Improvement," International Institute of Welding, Commission VIII Document 197-64, 1964.

248. Gadzikiewicz, I., and Dominiczak, K., "Histological Studies of Certain Parenchymal Organs of the Guinea Pig Exposed to the Action of Welding Fumes Using Electrodes of the EP47-28P Type," Rocz. Pomor. Akad. Med., 14, 171-175 (1968).

249. Senczuk, W., Mlynarczyk, M., and Mlynarczyk, W., "Effect of Toxic Dusts on Sulfur Amino Acids and their Metabolism," Ann. Pharm., 8, 121-129 (1970).

250. Senczuk, W., Litewka, B., Orlowski, J., and Kaczor, M., "Glycoproteins of Blood Serum of Animals following Exposure to Toxic Dusts," Ann. Pharm., 8, 111-119 (1970).

251. Mahler, H. R., and Cordes, E. H., "Biological Chemistry," Harper & Row, New York (1966).

252. Kukula, P., Trzeszczynski, J., and Jachimowicz, I., "Manganese Content in Guinea Pigs Exposed to Dusts and Fumes," Rocz. Pomorskiej. Akad. Med. Szczecinie, 14, 191-197 (1968).

253. Kolanecki, L., and Skucinski, S., "Manganese Content in Brain and Pancreas of Animals Exposed to Welding Fumes," Rocz. Pomorskiej. Akad. Med. Szczecin, 14, 183-190 (1968).

254. Bunka, H., and Kiziewicz, L., "Silica Content in Guinea Pigs Exposed to Dusts and Fumes During Experimental Welding with EP 47-28P Electrodes," Rocz. Pomorskiej Akad. Med. Szczecinie, 14, 163-169 (1968).

255. Naumenko, I. M., and Frolov, A. F., "Combined Effect of an Arc Welding Aerosol and Influenza Virus on Rat Lung Tissue," Gig. Sanit. 33(5), 111-113 (1968).

256. Gardner, L. U., and McCrum, D. S., "Effects of Daily Exposure to Arc Welding Fumes and Gases upon Normal and Tuberculous Animals," J. Ind. Hygiene Toxicol., 24(7), 173-182 (1942).

257. Walther, G., and Szilagy, S., "Histochemical Evidence of an Early Stage of Retinal Damage by Electric Welding Arc in Rabbits," Int. Arch. Arbeitsmed., 26(3), 189-197 (1970).

258. Ham, W. T. Jr., Mueller, H. A., and Sliney, D. H., "Retinal Sensitivity to Damage from Short Wavelength Light," Nature, 260, 153-155 (1976).

259. Petrilli, F. L., and de Flora, S., "Toxicity and Mutagenicity of Hexavalent Chromium on Salmonella Typhimurium," Appl. Environ. Microbiol., 33(4), 805-809 (1977).

260. Fenton, E. A., "Welding," in: "Kirk-Othmer Encyclopedia of Chemical Technology," 2nd ed., Vol. 22, A. Standen, ed., John Wiley & Sons, NY., p. 241-252.

261. American Society of Metals, "Metals Handbook," 8th ed., Vol. 6, American Society of Metals, Metals Park, OH (1971).

262. Steel, J., and Anderson, J. T., "Toxic Constituents of Welding Fumes," Ann. Occup. Hyg., 9(3), 103-111 (1966).

263. Hummitzsch, W., "Gas and Fumes Evolved during the Welding of Ferrous Metals," Werkstatt u. Betrief, 88(6), 313-316 (1955).

264. Thrysin, E., Gerhardsson, G., and Forssman, S., "Fumes and Gases in Arc Welding," Arch. Ind. Hyg. Occup. Med., 6(5), 381-402 (1952).

265. Tebbens, B. D., and Drinker, P., "Ventilation in Arc Welding with Coated Electrodes," J. Ind. Hyg. Toxicol., 23, 322 (1941).

266. Kobayashi, M., Maki, S., and Ohe, I., "Factors Affecting the Amount of Fumes Generated by Manual Metal Arc Welding," International Institute of Welding, Commission VIII Document 670-76, 1976.

267. Silverman, L., and Gilbert, H., "Working Conditions Ambient to Inert-Gas Shielded Metal Arc Welding," Weld. J., 33(5), 218s-229s (1954).

268. Phillips, A., ed., "Welding Handbook," Section 2, 6th ed., American Welding Society, Miami, FL.

269. Breslin, A. J., and Harris, W. B., "Use of Thoriated Tungsten Electrodes in Inert Gas Shielded Arc Welding," Am. Ind. Hyg. Assoc. Quart., 13, 191-195 (1952).

270. Breslin, A. J., "Potential Hazard from Use of Thoriated Tungsten Electrodes," Weld. J., 32, 134 (1953).

271. Entwistle, H., "Welding, Thermal Cutting," in: "Occupational Health and Safety," Vol. II, McGraw-Hill Book Co., NY (1972), p. 1488-1494.

272. Byczkowski, S., Bohdanowicz, A., Kopczynski, W., and Senczuk, W., "Evaluation of Working Conditions of Welders using Submerged Arc Welding," International Institute of Welding, Commission VIII Document 196-64, 1964.

273. Pantucek, M., "Influence of Filler Materials on Air Contamination in Manual Electric Arc Welding," Am. Ind. Hyg. Assoc. J., 32(10), 687-692 (1971).

274. Hartford, W. H., and Copson, R. L., "Chromium Compounds," in: "Kirk-Othmer Encyclopedia of Chemical Technology," 2nd ed., Vol. 5, A. Standen, ed., John Wiley & Sons, Inc., NY, p. 473-516.

275. Patty, F. A., ed., "Industrial Hygiene and Toxicology," 2nd ed., Vol. II, John Wiley & Sons, Inc., NY (1963).

276. Fairhall, L. T., "Industrial Toxicology," The Williams and Wilkins Co., Baltimore, MD (1957), 376p.

277. Edwards, J. O., "Ozone in Inorganic Chemistry," in: "Ozone Chemistry and Technology," J. Murphy and J. Orr, eds., Franklin Institute Press, Philadelphia, PA (1975), p. 187-190.

278. Ferry, J. J., and Ginther, G. B., "Gases Produced by Inert Arc Welding," Weld. J., 32, 396-398 (1953).

279. Fannick, N. L., and Corn, M., "The Plasma Jet: Industrial Hygiene Aspects and a Survey of Current U.S. Practices for Employee Protection," Am. Ind. Hyg. Assoc. J., 30(3), 226-235 (1969).

280. Elkins, H. B., "Nitrogen Dioxide – Rate of Oxidation of Nitric Oxide and its Bearing on the Nitrogen Dioxide Content of Electric Arc Fumes," J. Ind. Hyg. Toxicol., 28, 37-39 (1946).

281. Fay, H., Mohr, P. H., and McDaniel, P. W., "Nitrogen Dioxide and Ozone Concentrations in Welding Operations," Am. Ind. Hyg. Assoc. Quart., 18, 19-28 (1957).

282. Hummitzsch, W., "Permissible Gas Concentrations During CO_2 Welding," Schweissen u. Schneiden, 12(8), 274-275 (1960).

283. Erman, M., Raiskiy, E., and Potapevskiy, A., "Hygienic Assessment of CO_2 Welding in Shipyards," Automatic Weld., 21(4), 63-67 (1968).

284. Dahlberg, J. A., Andersson, H., and Wettstrom, R., "On the Decomposition of Trichloroethylene, Perchloroethylene and Methyl Chloroform in Welding Works," International Institute of Welding, Commission VIII Document 585-74, 1974.

285. Dahlberg, J. A., Christiansen, V. O., and Eriksson, E. A., "On the Formation of Phosgene by Photo-Oxidation of Methyl Chloroform in Welding," Ann. Occup. Hyg., 16, 41-46 (1973).

286. Mantani, T., Yabuta, T., and Notani, H., "On the Formation of Phosgene in Welding Works," International Institute of Welding, Commission VIII Document 420-70, 1970.

287. Dahlberg, J. A., and Myrin, L. M., "The Formation of Dichloroacetyl Chloride and Phosgene from Trichloroethylene in the Atmosphere of Welding Shops," Ann. Occup. Hyg., 14, 269-274 (1971).

288. Andersson, H. F., Dahlberg, J. A., and Wettstrom, R., "Phosgene Formation during Welding in Air Contaminated with Perchloroethylene," Ann. Occup. Hyg., 18(2), 129-132 (1975).

289. Huebner, H. J., Krause, E., Ruge, J., and Sutter, E., "The Measurement of Radiation from Arc Welding Carried out by Means of Various Welding Processes (Including Plasma Cutting). A Contribution Towards the Revision of German Industrial Standard Din 4647 Sheet 1: The Use of Optical Glass in Eye-Protection Equipment; Protective Filter Glass for Welding," Schweissen u. Schneiden 24(8), 290-293 (1974), translated into English in National Research Council of Canada Report No. NRC-TT-1779.

290. Van Someren, E., and Rollason, E. C., "Radiation from the Welding Arc. Its Effect on the Eye," Weld. J., 27, 448s-452s (1948).

291. Glickstein, S. S., "Temperature Measurements in a Free Burning Arc," Weld. J., 55(8), 222s-229s (1976).

292. Marshall, W. J., Sliney, D. H., Lyon, T. L., Krial, N. P., and Del Valle, P. F., "Nonionizing Radiation Protection Special Study No. 42-0312-77, Evaluation of the Potential Retinal Hazards from Optical Radiation Generated by Electric Welding and Cutting Arcs," National Technical Information Service Report AD-A043023, 1977.

293. Dahlberg, J. A., "The Intensity and Spectral Distribution of Ultraviolet Emission from Welding Arcs in Relation to the Photodecomposition of Gases," Ann. Occup. Hyg., 14, 259-267 (1971).

294. Mechev, V. S., and Eroshenko, L. E., "Research into the Spectrum of Radiation by the Argon-Shielded Arc Close to the Electrodes," Automatic Weld., 25(8), 1-5 (1972).

295. Horstman, S. W., Emmett, E. A., and Kreichelt, T. E., "Field Study of Potential Ultraviolet Exposures from Arc Welding," Weld. J., 55(5), 121s-126s (1976).

296. Zaborski, L., "Studies on the Luminance of Welding Arcs," Bull. Inst. Marit. Trop. Med. Gdynia, 28(3/4), 129-134 (1977).

297. U.S. National Institute of Occupational Safety and Health, "Criteria for a Recommended Standard. Occupational Exposure to Hot Environments," NIOSH.

298. Weiss, W., "Cigarette Smoking and Diffuse Pulmonary Fibrosis," Am. Rev. Resp. Dis. 99, 67-72 (1969).

299. Papp, J. P., "Metal Fume Fever," Postgrad. Med., 43(3), 160-163 (1968).

300. Felczak, J., "Pneumoconiosis of Electric Welders and Experiments on the Influence of Welding Fumes on the Pulmonary Tissue of Rats," Pol. Przegl. Radiol., 31(6), 803-816 (1967).

301. Clark, B., "Welding Filters and Thermal Damage to the Retina," National Technical Information Service Report AD 674 960, 1968.

302. Ishimi, A., Ohmoto, M., "Toxicity of Electric Arc Welding Fumes. II. Distribution of Iron, Manganese, and Magnesium in Rats Exposed to Arc Welding Fumes," Sangyo Igaku, 16(2), 113-123.

303. Leslie, A. C., Winchester, J. W., Leysieffer, F. W., and Ahlberg, M. S., "Prediction of Health Effects of Pollution Aerosols," Trace Subst. Environ. Health, 10, 497-504 (1976).

304. Hewitt, P. J., and Hicks, R., "Neutron Activation Analysis of Blood and Body Tissue from Rats Exposed to Welding Fumes," International Institute of Welding, Commission VIII Document 730-77, 1977.

305. Swensson, A., "Tissue Reaction to Different Types of Amorphous Silica," In: C.N. Davies, ed., Inhaled Particles and Vapors II, Pergamon Press (1967).

306. Eiso, K. A., "On the Clinical Peculiarities of Manganese Poisoning in Electric Welders," Gig. Tr. Prof. Zabol., 10, 39-41 (1966).

307. 29 Code of Federal Regulations 1910.1000 (1976).

308. 29 Code of Federal Regulations 1910.95 (1976).